_____ *Technologies of Freedom*

Technologies
of Freedom

Ithiel de Sola Pool

The Belknap Press of
Harvard University Press
Cambridge, Massachusetts
and London, England

Library of Congress Cataloging in Publication Data

Pool, Ithiel de Sola, 1917–
     Technologies of freedom.

     Includes index.
     1. Mass media—Law and legislation—United States.   2. Tele-
communication—Law and legislation—United States.   3. Liberty of
speech—United States.   4. Liberty of the press—United
States.   5. Technology—Social aspects—United States.   I. Title.
KF2750.P66   1983   343.73′0998   82-24498
ISBN 0-674-87232-0 (cloth)
ISBN 0-674-87233-9 (paper)

For Jean, Jonathan, Jeremy, and Adam

Acknowledgments

Support for the research for this book came mainly from the John and Mary Markle Foundation. The primary environment for its writing was the Research Program on Communications Policy at the Massachusetts Institute of Technology. But parts of it were written at other hospitable places of retreat and stimulation: Churchill College, the University of Cambridge; the Long-Range Planning Units of British Telecoms, Cambridge; the Institute for Communications Research, Keio University; the Research Institute of Telecommunications and Economics, Tokyo; and the East-West Center and the University of Hawaii Law Library, Honolulu. Portions of the book were produced as Phi Beta Kappa lectures. The material in Chapter 6 dealing with events leading up to the Radio Act of 1927 and the Communications Act of 1934 was done with the support of Project '87 of the American Political Science Association and was written jointly with Priscilla Fox. Parts of Chapter 8 appeared in different form in my article "Print Culture and Video Culture" in the Fall 1982 issue of *Daedalus*.

The people whose ideas I have borrowed and who have critiqued parts of the manuscript are far more numerous than I know, but those from whom I am particularly aware of learning include Charles Jackson, Henry Geller, Marcus Cohn, Clay T. Whitehead, Burton Nadler, Alex Reid, Russell Neuman, and Charles Jonscher. On copyright I have drawn ideas from my work with Richard Solomon and the late Harry Bloom, and on soft technological determinism, from the work of John Thomson.

The manuscript has greatly benefited from the good ideas and meticulous attention of Michael Aronson and Virginia LaPlante of Harvard University Press. The greatest burden has been borne by Nancy Wilson and, above all, by Suzanne Planchon, who tirelessly managed the manuscript on every step of its way.

Contents

_____ *Technologies of Freedom*

1. A Shadow Darkens

Civil liberty functions today in a changing technological context. For five hundred years a struggle was fought, and in a few countries won, for the right of people to speak and print freely, unlicensed, uncensored, and uncontrolled. But new technologies of electronic communication may now relegate old and freed media such as pamphlets, platforms, and periodicals to a corner of the public forum. Electronic modes of communication that enjoy lesser rights are moving to center stage. The new communication technologies have not inherited all the legal immunities that were won for the old. When wires, radio waves, satellites, and computers became major vehicles of discourse, regulation seemed to be a technical necessity. And so, as speech increasingly flows over those electronic media, the five-century growth of an unabridged right of citizens to speak without controls may be endangered.

Alarm over this trend is common, though understanding of it is rare. In 1980 the chairman of the American Federal Communications Commission (FCC) sent a shiver through print journalists when he raised the question of whether a newspaper delivered by teletext is an extension of print and thus as free as any other newspaper, or whether it is a broadcast and thus under the control of the government.[1] A reporter, discussing computerized information services, broached an issue with far-reaching implications for society when she asked, "Will traditional First Amendment freedom of the press apply to the signals sent out over telephone wires or television cables?"[2] William S. Paley, chairman of the Columbia Broadcasting System (CBS), warned the press: "Broadcasters and print people have been so busy improving and defining their own turf that it has escaped some of us how much we are being drawn together by the vast revolution in 'electronification' that is changing the face of the media today . . . Convergence of delivery mechanisms for news and information raises anew some critical First Amendment questions

. . . Once the print media comes into the home through the television set, or an attachment, with an impact and basic content similar to that which the broadcasters now deliver, then the question of government regulation becomes paramount for print as well."³ And Senator Bob Packwood proposed a new amendment to the Constitution extending First Amendment rights to the electronic media, on the assumption that they are not covered now.

Although the first principle of communications law in the United States is the guarantee of freedom in the First Amendment, in fact this country has a trifurcated communications system. In three domains of communication—print, common carriage, and broadcasting—the law has evolved separately, and in each domain with but modest relation to the others.

In the domain of print and other means of communication that existed in the formative days of the nation, such as pulpits, periodicals, and public meetings, the First Amendment truly governs. In well over one hundred cases dealing with publishing, canvassing, public speeches, and associations, the Supreme Court has applied the First Amendment to the media that existed in the eighteenth century.

In the domain of common carriers, which includes the telephone, the telegraph, the postal system, and now some computer networks, a different set of policies has been applied, designed above all to ensure universal service and fair access by the public to the facilities of the carrier. That right of access is what defines a common carrier: it is obligated to serve all on equal terms without discrimination.

Finally, in the domain of broadcasting, Congress and the courts have established a highly regulated regime, very different from that of print. On the grounds of a supposed scarcity of usable frequencies in the radio spectrum, broadcasters are selected by the government for merit in its eyes, assigned a slice each of the spectrum of frequencies, and required to use that assignment fairly and for community welfare as defined by state authorities. The principles of common carriage and of the First Amendment have been applied to broadcasting in only atrophied form. For broadcasting, a politically managed system has been invented.

The electronic modes of twentieth century communication, whether they be carriers or broadcasters, have lost a large part of the eighteenth and nineteenth century constitutional protections of no prior restraint, no licenses, no special taxes, no regulations, and no laws. Every radio spectrum user, for example, must be licensed. This requirement started in 1912, almost a decade before the beginning of broadcasting, at a time when radio was used mainly for maritime communication. Because the United States Navy's communications were suffering interference, Congress, in an effort at remedy, imposed licensing on transmitters, thereby breaching a tradition that went back to John Milton against requiring licenses for communicating.

Regulation as a response to perceived technical problems has now reached the point where transmissions enclosed in wires or cables, and therefore causing no over-the-air interference, are also licensed and regulated. The FCC claims the right to control which broadcast stations a cablecaster may or must carry. Until the courts blew the whistle, the rules even barred a pay channel from performing movies that were more than three or less than ten years old. Telephone bills are taxed. A public network interconnecting computers must be licensed and, according to present interpretations of the 1934 Communications Act, may be denied a license if the government does not believe that it serves "the public convenience, interest, or necessity."

Both civil libertarians and free marketers are perturbed at the expanding scope of communications regulation. After computers became linked by communications networks, for example, the FCC spent several years figuring out how to avoid regulating the computer industry. The line of reasoning behind this laudable self-restraint, known as deregulation, has nothing to do, however, with freedom of speech. Deregulation, whatever its economic merits, is something much less than the First Amendment. The Constitution, in Article 1, Section 8, gives the federal government the right to regulate interstate commerce, but in the First Amendment, equally explicitly, it excludes one kind of commerce, namely communication, from government authority. Yet here is the FCC trying to figure out how it can avoid regulating the commerce of the computer industry (an authority Congress could have given, but never did) while continuing to regulate communications whenever it considers this necessary. The Constitution has been turned on its head.

The mystery is how the clear intent of the Constitution, so well and strictly enforced in the domain of print, has been so neglected in the electronic revolution. The answer lies partly in changes in the prevailing concerns and historical circumstances from the time of the founding fathers to the world of today; but it lies at least as much in the failure of Congress and the courts to understand the character of

the new technologies. Judges and legislators have tried to fit technological innovations under conventional legal concepts. The errors of
understanding by these scientific laymen, though honest, have been
mammoth. They have sought to guide toward good purposes technologies they did not comprehend.

"It would seem," wrote Alexis de Tocqueville, "that if despotism
were to be established among the democratic nations of our days . . .
it would be more extensive and more mild; it would degrade men
without tormenting them." This is the kind of mild but degrading
erosion of freedom that our system of communication faces today,
not a rise of dictators or totalitarian movements. The threat in
America, as Tocqueville perceived, is from well-intentioned policies,
with results that are poorly foreseen. The danger is "tutelary
power," which aims at the happiness of the people but also seeks to
be the "arbiter of that happiness."[4]

Yet in a century and a half since Tocqueville wrote, the mild despotism that he saw in the wings of American politics has not become
a reality. For all his understanding of the American political system,
he missed one vital element of the picture. In the tension between
tutelary and libertarian impulses that is built into American culture,
a strong institutional dike has held back assaults on freedom. It is the
first ten amendments to the Constitution. Extraordinary as this may
seem, in Tocqueville's great two volumes there is nowhere a mention
of the Bill of Rights!

The erosion of traditional freedoms that has occurred as government has striven to cope with problems of new communications
media would not have surprised Tocqueville, for it is a story of how,
in pursuit of the public good, a growing structure of controls has
been imposed. But one part of the story would have surprised him,
for it tells how a legal institution that he overlooked, namely the First
Amendment, has up to now maintained the freedom and individu-
alism that he saw as endangered.

A hundred and fifty years from now, today's fears about the future of free expression may prove as alarmist as Tocqueville's did.
But there is reason to suspect that our situation is more ominous.
What has changed in twentieth century communications is its technological base. Tocqueville wrote in a pluralistic society of small enterprises where the then new mass media consisted entirely of the
printed press which the First Amendment protected. In the period
since his day, new and mostly electronic media have proliferated in

the form of great oligopolistic networks of common carriers and broadcasters. Regulation was a natural response. Fortunately and strangely, as electronics advances further, another reversal is now taking place, toward growing decentralization and toward fragmentation of the audience of the newest media. The transitional era of giant media may nonetheless leave a permanent set of regulatory practices implanted on a system that is coming to have technical characteristics that would otherwise be conducive to freedom.

The causal relationships between technology and culture are a matter that social scientists have long debated. Some may question how far technological trends shape the political freedom or control under which communication takes place, believing, as does Daniel Bell, that each subsystem of society, such as techno-economics, polity, and culture, has its own heritage and axial principles and goes its own way.[5] Others contend, like Karl Marx or Ruth Benedict, that a deep commonality binds all aspects of a culture. Some argue that technology is neutral, used as the culture demands; others that the technology of the medium controls the message.

The interaction over the past two centuries between the changing technologies of communication and the practice of free speech, I would argue, fits a pattern that is sometimes described as "soft technological determinism." Freedom is fostered when the means of communication are dispersed, decentralized, and easily available, as are printing presses or microcomputers. Central control is more likely when the means of communication are concentrated, monopolized, and scarce, as are great networks. But the relationship between technology and institutions is not simple or unidirectional, nor are the effects immediate. Institutions that evolve in response to one technological environment persist and to some degree are later imposed on what may be a changed technology. The First Amendment came out of a pluralistic world of small communicators, but it shaped the present treatment of great national networks. Later on, systems of regulation that emerged for national common carriers and for the use of "scarce" spectrum for broadcasting tended to be imposed on more recent generations of electronic technologies that no longer require them.

Simple versions of technological determinism fail to take account of the differences in the way things happen at different stages in the life cycle of a technology. When a new invention is made, such as the telephone or radio, its fundamental laws are usually not well understood. It is designed to suit institutions that already exist, but in its early stages if it is to be used at all, it must be used in whatever form it proved experimentally to work. Institutions for its use are thus designed around a technologically determined model. Later, when scientists have comprehended the fundamental theory, the early technological embodiment becomes simply a special case. Alternative devices can then be designed to meet human needs. Technology no longer need control. A 1920s motion picture had to be black and white, silent, pantomimic, and shown in a place of public assembly; there was no practical choice. A 1980s video can have whatever colors, sounds, and three-dimensional or synthetic effects are wanted, and can be seen in whatever location is desired. In the meantime, however, an industry has established studios, theaters, career lines, unions, funding, and advertising practices, all designed to use the technology that is in place. Change occurs, but the established institutions are a constraint on its direction and pace.

Today, in an era of advanced (and still advancing) electronic theory, it has become possible to build virtually any kind of communications device that one might wish, though at a price. The market, not technology, sets most limits. For example, technology no longer imposes licensing and government regulation. That pattern was established for electronic media a half-century ago, when there seemed to be no alternative, but the institutions of control then adopted persist. That is why today's alarms could turn out to be more portentous than Tocqueville's.

The key technological change, at the root of the social changes, is that communication, other than conversation face to face, is becoming overwhelmingly electronic. Not only is electronic communication growing faster than traditional media of publishing, but also the convergence of modes of delivery is bringing the press, journals, and books into the electronic world. One question raised by these changes is whether some social features are inherent in the electronic character of the emerging media. Is television the model of the future? Are electromagnetic pulses simply an alternative conduit to deliver whatever is wanted, or are there aspects of electronic technology that make it different from print—more centralized or more decentralized, more banal or more profound, more private or more government dependent?

The electronic transformation of the media occurs not in a vacuum but in a specific historical and legal context. Freedom for publishing has been one of America's proudest traditions. But just what is it that the courts have protected, and how does this differ from how the courts acted later when the media through which ideas flowed came to be the telegraph, telephone, television, or computers? What images did policy makers have of how each of these media works; how far were their images valid; and what happened to their images when the facts changed?

In each of the three parts of the American communications system—print, common carriers, and broadcasting—the law has rested on a perception of technology that is sometimes accurate, often inaccurate, and which changes slowly as technology changes fast. Each new advance in the technology of communications disturbs a status quo. It meets resistance from those whose dominance it threatens, but if useful, it begins to be adopted. Initially, because it is new and a full scientific mastery of the options is not yet at hand, the invention comes into use in a rather clumsy form. Technical laymen, such as judges, perceive the new technology in that early, clumsy form, which then becomes their image of its nature, possibilities, and use. This perception is an incubus on later understanding.

The courts and regulatory agencies in the American system (or other authorities elsewhere) enter as arbiters of the conflicts among entrepreneurs, interest groups, and political organizations battling for control of the new technology. The arbiters, applying familiar analogies from the past to their lay image of the new technology, create a partly old, partly new structure of rights and obligations. The telegraph was analogized to railroads, the telephone to the telegraph, and cable television to broadcasting. The legal system thus invented for each new technology may in some instances, like the First Amendment, be a *tour de force* of political creativity, but in other instances it may be less worthy. The system created can turn out to be inappropriate to more habile forms of the technology which gradually emerge as the technology progresses. This is when problems arise, as they are arising so acutely today.

Historically, the various media that are now converging have been differently organized and differently treated under the law. The outcome to be feared is that communications in the future may be unnecessarily regulated under the unfree tradition of law that has been applied so far to the electronic media. The clash between the print, common carrier, and broadcast models is likely to be a vehement communications policy issue in the next decades. Convergence of modes is upsetting the trifurcated system developed over the past two hundred years, and questions that had seemed to be settled centuries ago are being reopened, unfortunately sometimes not in a libertarian way.

The problem is worldwide. What is true for the United States is true, *mutatis mutandis*, for all free nations. All have the same three systems. All are in their way deferential to private publishing but allow government control or ownership of carriers and broadcasters. And all are moving into the era of electronic communication. So they face the same prospect of either freeing up their electronic media or else finding their major means of communication slipping back under political control.

The American case is unique only in the specific feature of the First Amendment and in the role of the courts in upholding it. The First Amendment, as interpreted by the courts, provides an anchor for freedom of the press and thus accentuates the difference between publishing and the electronic domain. Because of the unique power of the American courts, the issue in the United States unfolds largely in judicial decisions. But the same dilemmas and trends could be illustrated by citing declarations of policy and institutional structures in each advanced country.

If the boundaries between publishing, broadcasting, cable television, and the telephone network are indeed broken in the coming decades, then communications policies in all advanced countries must address the issue of which of the three models will dominate public policy regarding them. Public interest regulation could begin to extend over the print media as those media increasingly use regulated electronic channels. Conversely, concern for the traditional notion of a free press could lead to finding ways to free the electronic media from regulation. The policies adopted, even among free nations, will differ, though with much in common. The problems in all of them are very much the same.

The phrase "communications policy" rings oddly in a discussion of freedom from government. But freedom is also a policy. The question it poses is how to reduce the public control of communications in an electronic era. A policy of freedom aims at pluralism of expression rather than at dissemination of preferred ideas.

A communications policy, or indeed any policy, may be mapped on a few central topics:

Definition of the domain in which the policy operates
Availability of resources
Organization of access to resources
Establishment and enforcement of norms and controls
Problems at the system boundaries

The *definition of the domain* of a communications policy requires consideration of the point at which human intercourse becomes something more than communication. In American law, at some point speech becomes action and thus no longer receives First Amendment protection. Similar issues arise as to whether under the law pornography is speech, and whether commercial speech is speech.

The *availability of resources* raises another set of questions. Tools, money, raw materials, and labor are all required in order to carry on communication. The press needs newsprint; broadcasters need spectrum. How much of these can be made available, and at what prices can they be had?

The *organization of access to these resources* can be by a market or by rationing by the state. There may be a diversity of sources of resources, or there may be a monopoly. How much freedom is allowed to those who control the resources to exercise choice about who gets what? Are resources taxed, are they subsidized, or neither? How is intellectual property defined, and how is it protected?

The *exercise of regulation and control* over communication is a central concern in any treatise on freedom. How much control are policy makers allowed to exercise? What are the limitations on them, such as the First Amendment? May they censor? May they license those who seek to communicate? What norms control the things that communicators may say to each other? What is libel, what is slander, what violates privacy or security? Who is chosen to enforce these rules, and how?

The *problems encountered at the system boundaries* include the transborder issues that arise when there is a conflict of laws about communications which cross frontiers. Censorship is often imposed for reasons of national security, cultural protection, and trade advantage. These issues, which have not been central in past First Amendment controversies, are likely to be of growing importance in the electronic era.

From this map of policy analysis can be extracted what social scientists sometimes call a mapping sentence, a brief but formal statement of the problem to be analyzed in this book. It seeks to understand the impact of resource availability, as affected both by technology and by the organization of access to the resources, upon freedom from regulation and control. The specific question to be answered is whether the electronic resources for communication can be as free of public regulation in the future as the platform and printing press have been in the past. Not a decade goes by in free countries, and not a day in the world, without grim oppressions that bring protesters once more to picket lines and demonstrations. Vigilance that must be so eternal becomes routine, and citizens grow callous.

The issue of the handling of the electronic media is the salient free speech problem for this decade, at least as important for this moment as were the last generation's issues for them, and as the next generation's will be for them too. But perhaps it is more than that. The move to electronic communication may be a turning point that history will remember. Just as in seventeenth and eighteenth century Great Britain and America a few tracts and acts set precedents for print by which we live today, so what we think and do today may frame the information system for a substantial period in the future.

In that future society the norms that govern information and communication will be even more crucial than in the past. Those who read, wrote, and published in the seventeenth and eighteenth centuries, and who shaped whatever heritage of art, literature, science, and history continues to matter today, were part of a small minority, well under one-tenth of the work force. Information activities now occupy the lives of much of the population. In advanced societies about half the work force are information processors.[6] It would be dire if the laws we make today governing the dominant mode of information handling in such an information society were subversive of its freedom. The onus is on us to determine whether free societies in the twenty-first century will conduct electronic communication under the conditions of freedom established for the domain of print through centuries of struggle, or whether that great achievement will become lost in a confusion about new technologies.

2. Printing and the Evolution of a Free Press

When the American Bill of Rights was framed, free speech seemed to be a clear and well-defined notion, one that grew out of the religious controversies of the previous centuries and, more recently, out of the American Revolution. People had convictions that they shared with fellow congregants, but which often differed from those of their neighbors or rulers. Where they were blessed with free speech, they could express those beliefs from pulpits and in face-to-face discussions.

The derivative notion of freedom of the press was not very different. A print shop, like a pulpit, gave voice to an individual. Often the printer was also a publisher, with perhaps just an employee or two. Freedom of the press, like free speech, basically meant that individuals could express themselves. Obtaining access to the needed resources was no strain on ingenuity. All that government had to do was keep hands off; if it did so, motivated individuals would be able to publish their views.

From the beginning some differences distinguished freedom of the press from freedom of speech. The physical printing plant was potentially hostage to state action, so the printer was sensitive to discriminatory taxes or other harassments. Copyright was important, as the publisher relied on sales to support costs that did not exist for the face-to-face preacher. And the publisher could be helped greatly by postal subsidies or public advertisements. So the print model that emerged, though equated with freedom of speech, was like it only in part. It was like it in excluding any regulatory action by the government, but different in accepting some minimal, preferably arm's-length public support through copyright and the mails. Partisans of a free press prefer not to draw attention to even that limited state role, but it was there.

In one important respect the original imagery behind the print model has ceased to reflect reality. Publishing is rarely now the ex-

pression of just an individual. It is undertaken by large organizations. In the United States, a few wire services serve one-to-a-city
newspapers. In other democratic countries too, though the trend has
not gone as far, the number of competitive newspapers is declining.
Yet in most advanced countries, despite the growth in large-scale
publishing, freedom of publishing remains intact.

The Technology of Print

The technology that has led to the Western system of a free press
began in China, when paper from textiles was invented early in the
second century and spread westward. By the twelfth century, paper
was manufactured in Spain, and by the thirteenth in Italy. Around
1300 the price of the new medium was but one-sixth the cost of
parchment.[1]

The invention of movable type, attributed to one Pi Sheng in
1041–1049, did not become important in imperial China because for
what that society wanted, frequent reprints of a relatively small
number of classics, printing from full-page blocks was a good solution. In Korea, however, for more general uses, movable type was
adopted and developed. In 1241 metal type was substituted for
earthenware characters, and in 1403 a Korean king proclaimed, "It is
our will and law that type shall be produced from copper and that
various books be printed, so that in this way knowledge may be
more widely disseminated for the countless needs of all."[2] Yet in
neither China nor Korea did an explosion of private publishing take
place, nor did a tradition of freedom stem from printing, for printing
was used as the monarch's mandarins directed.

Only later, in the different context of the unique history of the
modern West, did printing become a technology of freedom. Issues
about rights of individuals were, however, not unique to the modern
West. On the contrary, some notion of a right to speak is universal,
just as is repression. There are in every culture rules, evidenced in
ethnographic reports, on who may speak up about what. In traditional Samoa, for example, a chief made no major decision until his
"talking chief," at the next level down, had quietly consulted the
populace to seek a consensus. Ordinary citizens could not discuss
issues in public, nor even allow certain words that referred to politics
to cross their lips. Words alluding to power were the Samoans' dirty
words, not to be spoken by any but chiefs. When the talking chiefs

found consensus, they announced it from the platform, and the chief listened and acquiesced.[3]

Such were the rules in one quite authoritarian society. In other councils of tribes or villages, such as Indian panchayats, a range of practices are found about who may state a view. Most often such rules were elitist and restrictive, but occasionally they were freer and contentious.[4] Particularly where religious inspiration gave unordained people the conviction that through their tongues gods or demons spoke, all sorts of views might come to be expressed.[5]

For the origins of the particular Western tradition of free speech, however, one looks to the shores of the Mediterranean. Jewish prophets exposing iniquity and Athenian Sophists training students to debate provide precedents for two millennia of advocates and dissidents in the West. Although the Jews and the Greeks were literate, and some of what they transcribed later became our great books, the ongoing debates in their day were oral. The prophets exhorted congregants, and the orators taught by the Sophists declaimed in the agora. Neither the rebelliousness of the prophets nor the litigiousness of the orators, however, became the consistent practice of Western civilization. Empires, tyrannies, marauders, feudalism, and *auto-da-fés* make up more of Western history. Yet a heritage of speech, debate, science, religious doctrine, literary creation, and philosophical dispute was present in the seams of the Western world when in the middle of the fifteenth century the Chinese inventions finally came West.

Printing in the West took off in a different way from printing in China, both in its technology and in its use. The flat Korean squares of type, roughly the shape of an anagram piece, could not be firmly bound together in a frame. Gutenberg added a half-inch of base, which permitted the composed page of type to be held together and handled as a solid block that could be inked and have paper pressed to it. This system spread rapidly. By the 1490s the major states each had one or more important publishing centers. Between 1481 and 1501, 268 printers in Venice turned out two million volumes.[6]

As with many innovations, the full impact of printing could not be seen at once. Paul Lazersfeld once jokingly suggested that if a foundation had made a grant to researchers to evaluate printing a decade or two after its invention, they would have concluded that the new device was vastly overrated. Scribes were already producing the important books efficiently, and the new printers produced mainly the same old texts, such as the Bible, which were readily available to the tiny minority who were literate.[7] Hand copying continued to be competitively viable throughout the century. In the region of Paris and Orleans alone about 10,000 scribes were at work.[8] Often they even copied printed books, for when the typical edition of a few hundred copies of an early book ran out, it was more economical to meet the residual demand by hand.

In the end, however, printing had profound effects on society. The fact that a printer could produce on the average one volume a day, while a copyist produced two a year, made change inescapable.[9] Chief among the early social impacts of printing was the strengthening of Protestantism and the weakening of the Roman Catholic Church. When family Bibles became available to common people, the priest was no longer its exclusive interpreter. Tracts, sermons, and private interpretations of all sorts were diffused in print. Printing promoted Protestantism in less obvious ways too. Manuscript copying was one of the economic mainstays of the monasteries, whereas printing was done by bourgeois craftsmen. This displacement of jobs from the domain of the church to that of the guilds helped shift the balance of power.[10] Furthermore, commercial printers were motivated to publish more books for more profits. For this they combatted the church monopoly on imprimatur and, in pursuit of sales—just like today's publishers—solicited controversial manuscripts from all sects.[11]

Censorship and Control

The path from printing to liberalism was not a straight one. Repression is in fact most likely not before a technology of liberation comes along, but only afterward when the powers that be are challenged by the beginnings of change. In reaction to the heresies that flowed from the print shops, the Roman Catholic Church tightened censorship and controls.[12] Before printing, there had been no elaborate system of censorship and control over scribes. There did not have to be. The scribes were scattered, working on single manuscripts in monasteries. Moreover, single manuscripts rarely caused a general scandal or major controversy. There was little motive for central control, and control would have been impractical. But after printing, Pope Alexander VI issued a bull in 1501 against the unlicensed printing of books. In 1559 the *Index Expurgatorius* was first issued.

Printing by then was widespread enough to worry the authorities and centralized enough to present a target for control.

Governments reacted to the "menace" of the printed word. In Germany, censorship was introduced in 1529.[13] In 1557 the British crown, seeking to check seditious and heretical books, chartered the Stationers' Company and limited the right to print to the members of that guild. Thirty years later the Star Chamber, to curtail the "greate enormities and abuses" of "dyvers contentyous and disorderlye persons professinge the arte or mystere of Pryntinge or sellinge of bookes," restricted the right to print to the two universities and to the twenty-one existing shops in the city of London with their fifty-three presses.[14]

Countries that seriously restricted printing lost business to those that left it freer. British controls on type founding in 1637 made that country dependent on the Dutch for type. French repression on printing in the previous century, particularly the burning at the stake of the printer Etienne Dolet in 1546, caused many printers to flee to the Netherlands. A century later, Richelieu had to send to Holland for printers to open a royal printing plant. Such problems with authorities made printers radical and rebellious. Before the mob tore down the Bastille in 1789, over eight hundred authors, printers, and book dealers had been incarcerated there.[15]

So printing became a challenge to authority throughout Europe. The battle between publishers and censors led in England to press laws and institutions that would later shape American practices. Following the attempts to limit printing to the monopoly of the Stationers' Company, the British government tried three other types of control. One was to require licenses for printers to publish. A number of licensing acts were passed, the first of which was protested in 1644 by John Milton in a pamphlet that became the classic defense of free speech, the *Areopagitica*.

After the end of licensing in 1693, the principle strategy of press repression became taxation. From 1712, for a century and a half, the British government restricted the growth of a press it found obnoxious by raising its costs through taxes. *The Spectator*, for example, ceased publication in 1712 after the new tax made its circulation fall from 4000 to 1600.[16] Taxes were imposed on newsprint, on ads, and on newspapers themselves. The colonial protest against the Stamp Act in 1765 was addressed against just such a tax requiring the pasting of revenue stamps on printed documents.

Still another strategy of press control used by the British government was prosecution of its critics for criminal libel. Such suits were brought not by an injured private citizen but by the state itself for a reputed libel against the authorities.

American legislators and courts rejected these three abuses which publishing had suffered in their country of origin: licensing of the press, special taxes on the press, and prosecution for criminal libel. The unconstitutionality of licensing, which the American courts referred to as "previous" or "prior restraint," was decided as early as 1825.[17] The tradition against special taxes on the press, which British protestors such as Richard Cobden called "taxes on knowledge," was reaffirmed by the Supreme Court in 1936.[18] And the prohibition against criminal libel suits became an American tradition in the 1735 trial of Peter Zenger, accused of libeling the governor of New York. The colonial jury, disregarding the judge's instructions on the law, acquitted Zenger and thus made the law. Since 1964, libel suits brought by public officials or public figures against their critics, even when brought in their own capacity and not by the state, have been greatly restricted by the courts.[19]

The colonists' rejection of the various British attempts to impose government authority over the press were incorporated into the American Constitution by the First Amendment. This amendment creates a domain—of speech, religion, and press—in which the activities of private citizens shall be unregulated by government. "Congress," it says, "shall make no law . . . abridging freedom of speech or of the press."[20]

But the First Amendment is just one of three clauses in the Constitution that deal specifically with communications. Another is the copyright provision in Article 1, Section 8: "Congress shall have the power . . . To promote the Progress of Science and useful Arts, by securing for limited Times to Authors and Inventors the exclusive Right to their respective Writings and Discoveries." In Britain the practice of copyright, though not the word, began at the founding of the Stationers' Company when, for enforcement, the company was given the right to search for and seize anything printed contrary to statute or proclamation. Eight years later the company, under this power, created a system of copyright for its members. In 1709 the first copyright act for authors was passed by Parliament.[21]

The new notion of intellectual property represented by copyright was rooted in the technology of print. The printing press was a bottleneck where copies could be examined and controlled. In the passage from the author's pen to the reader's hand, the press was the logical place to apply controls, be it to censor sacrilege or sedition or to protect the author's intellectual property.

For modes of reproduction where such an easy locus of control as the printing press did not exist, the concept of copyright was not applied. It was not applied to conversation, or to speeches, or to the singing of songs whether in private or in public. Copyright was a specific adaption to a specific technology. The common law recognized this fact. The landmark case in the United States denied copyright protection to piano rolls because they were not "writings" in a tangible form readable by a human being.[22] This concept of copyright excluded from protection many new technologies of communication. But the motion picture industry, the recording industry, and more recently the broadcasting industry have all persuaded Congress to give them the protection that the courts proved unwilling to give.

The third provision in the Constitution dealing with communications gives Congress the power "To establish Post Offices and post Roads."[23] This provision put the federal government into the common carrier business. Only one of today's carrier systems then existed, the mails. A post office had been permanently established in Britain in 1656 and in the colonies in 1711.

Before that, the crown had farmed out grants and patents to private entrepreneurs to carry government correspondence. To make these franchises attractive, the franchisees were also allowed to carry letters for the general public for a fee, and others were forbidden to compete with the chosen carriers in doing so. This scheme for providing government with cheap communication was the origin of the postal monopoly. When governments started carrying the mails themselves, the monopoly principle was further reinforced.

After American independence, the fiscal tradition of the post as a source of revenue was retained, and so was the practice of monopoly. In the 1820s the balance of public policy shifted from one of subordinating the post office to the treasury department as a producer of revenue, to promoting it to a full-fledged department of government, consecrated above all to extending the benefits of development to remote parts of the country at a rapid pace.[24] Still another important as well as expensive social goal pursued through the post office in the nineteenth century was the diffusion of knowledge. Newspapers, and later books and magazines, were given large subsidies in mail rates.[25]

The constitutional injunctions to the federal government with regard to communications were thus in appearance somewhat contradictory, though in fact their goals were quite consistent. In one clause the government was told to keep its legislative hands off of speech and press, while in two others it was told to promote the conveyance of knowledge by means of copyright and postal service. But both the injunctions to restraint and the injunction to governmental activism had the common goal of facilitating autonomous communication by private individuals.

The Mass Media Revolution

The seeds of today's mass media were planted by Gutenberg and fertilized by constitutional concerns, but no mass media in a modern sense existed before the 1830s. Only then did the production and distribution of published words begin their remarkable growth, as a series of devices was invented for the cheap mass production of uniform messages. These inventions were a subset of the devices invented for mass production of commodities. The mass media revolution was a part of the industrial revolution, in which entrepreneurs found that they could drastically cut costs of production by using the factory system, power machinery, and assembly lines.[26] Standardized commodities poured out of great sheds in which engines drove machines and moved products down the assembly line past disciplined laborers each doing a small part. Hand crafts and cottage industries were displaced. The factory system made cheap goods available to consumers, albeit often uniform and prosaic ones.

Exactly the same concepts were applied, during the same era, to the production of mass media. The craftsman printer who produced one page at a time by hand for a total of two thousand sheets in a ten-hour day was displaced, first by the power press, adopted at the *London Times* in 1814, then by the rotary press, adopted at the *Times* in 1869, and finally by a series of similar inventions that introduced into the modern printing plant a power-driven assembly line like that in any other factory. These plants made it possible to produce penny newspapers, such as the *New York Sun* in 1833. Before then, five thousand copies had been a good circulation for an American newspaper. The *Sun* reached a circulation of twenty-seven thousand in two years, and its immediate successor, *The Herald*, reached forty thousand by 1836.

In the second half of the century, as the circulation of major papers reached six figures, newspapers became commercial enterprises more than organs of political tendency. Press magnates began putting out thicker papers, with fiction, sports, and features, which were designed to appeal to a mass audience regardless of the reader's ideology and were heavily dependent on income from advertising. A series of supplementary inventions made it possible for capital-intensive enterprises to collect, compose, and print news faster and thus publish more pages, and to deliver them throughout the metropolis faster. For example, telephones allowed reporters to dictate stories from the field instead of coming into the office. The linotype speeded the process of composition. Automotive trucks rushed papers to newsboys to deliver by dawn or hawk on the streets.

As in the Renaissance, developments in the manufacture and marketing of paper were as important as developments in printing itself. In 1804, to encourage the growth of the press, Congress repealed the customs duty on rags, the principle raw material for paper of the day. Through most of the nineteenth century, rag collectors plied city streets around the world to supply the booming demand created by mass publishing. The United States imported vast quantities of rags, Italy being the largest single supplier. The demand for paper in Britain was held down by taxes on publications. As the tax was not proportional to the size or value of the printed product, it was only a minor burden on a thick volume but a major burden on a thin newspaper or pamphlet. It thus served the government's purpose of permitting ponderous publications while discouraging the more popular and agitational ones. When taxes on printed publications were sharply reduced and then repealed, mass circulation started to boom in Britain too.[27] As a result, the demand for rags and the price of them soared around the world. The American press was in crisis in the 1860s, with the price of newspapers sometimes doubling.

The solution was the manufacture of paper from wood pulp instead of rags. In principle, it had long been understood that paper could be made from many kinds of cellulose-containing organic materials. In 1800, when Matthias Koops published a catalogue of substances that had been used from earliest times for conveying information, it included paper from straw, nettles, and wood.[28] But the effective implementation of these substitute processes required technical development which occurred only when rising prices provided an incentive. Using the new wood pulp paper, the press enjoyed unprecedented prosperity and growth.[29] Newspaper prices fell while their size grew. The *New York World*, which cost five cents in 1866, sold for two cents in 1882 although it had doubled in size in that year.[30]

The technologies and market developments that fostered the newspaper in the nineteenth century also helped magazines to proliferate. Monthlies or quarterlies could be distributed nationally by the mails, so second-class mail rates were particularly important to them. Before about 1870, subsidized postal delivery had been crucial to the economics of newspapers too, but eventually the mass commercial papers built their own metropolitan distribution networks, whereas national postal delivery remained vital to magazines. A reduced postage for books was established by Congress in 1852. Although book publishers opposed the government's extension of concessionary rates to magazines, their competitors, which they argued were meretricious in content and undeserving of federal aid, reduced rates were extended to magazines shortly thereafter.

The growth of magazines was aided by another market development, the emergence of brand name products.[31] To sell a product by means of brand name promotion requires a scale of manufacturing that can serve a large and even national market. It also requires transportation for economical delivery of the commodity throughout that market. It is facilitated by having a package on which the brand logo can be displayed. Before paper became cheap, the storekeeper scooped unlabeled butter, biscuits, or soap from a tub or bin. With low cost paper and color printing, the manufacturer could package these in conveniently sized units bearing a distinctive label. To promote these packaged products, use was made of billboards, newspapers, and magazines. The product had to be sold nationwide for the manufacturer to advertise in national magazines.

Thus Europe and America entered the mass media era by the vehicle of print; the printed word and mass communication were for a while synonymous. Then in the first quarter of the twentieth century nonprint mass media came into use. In 1889 Thomas Edison had made his first successful Kinetoscope for viewing moving pictures on film, which in 1895 the Lumieres in France and Woodville Latham in the United States combined with a projector. By 1905 the first motion picture theater was opened in Pittsburgh, Pennsylvania. Fifteen years later radio broadcasting began with the opening of station KDKA, also in Pittsburgh. By 1977 broadcasting had grown to the point where, according to a census of communications flows, average

Americans consumed four times as many words electronically as they read in print.[32] In the total flow of media-delivered information, the relative part carried by newspapers, magazines, and books has dropped from being virtually all of it to being only 18 percent of the words to which people expose themselves.

Still, publication of printed matter has not declined absolutely. The volume of printed matter has kept growing, though much more slowly than for nonprint media. From 1960 to 1977 the circulation of daily papers in the United States grew by 5 percent, the circulation of magazines by 25 percent, and the distribution of books annually by 75 percent. By the end of that period .8 newspapers were circulated per household per day, and 172 magazine copies and 21 books per household per year. This sounds like a healthy industry. Printed expressions of opinion are not about to disappear.

Yet there is another side to the story. Not only has the growth of reading been surpassed by the growth of audio and video media, but since the late 1960s the growth of reading has been markedly slowing down and may even be approaching a halt. The nonprint media are not just passing the print media, but are for the first time showing signs of displacing them in part. America by 1970 seems to have reached a turning point, at least in the words made available to the public (not so much the words actually read and heard by the public). The rate of growth in print words supplied fell sharply and in a few areas actually declined. From 1960 to 1968 the words published by newspapers grew at .7 percent per annum, but from 1968 to 1977 the rate of change was −.3 percent per annum. The words that magazine publishers supplied to their audience increased at an annual rate of of 3 percent from 1960 to 1970, but only .1 percent from 1970 to 1977.

Projection of trends into the future is risky, but the sharp break in the late 1960s or early 1970s in what had been a fairly steady, if moderate, rate of growth in printed material arrests attention. Until recently, the new electronic media added something to the flow of communications without at the same time weakening the traditional print media. This seems no longer true.

The American figures and findings are not identical with what is found in other advanced industrialized countries, but the main patterns have much in common. Print publishing elsewhere remains a massive and even growing enterprise. Elsewhere too the electronic media have, however, been growing far faster and are becoming the dominant source of information. In most other advanced countries the phenomenon just seen in the United States of an abrupt check to the growth of print media has not yet occurred. But the volume of printed matter supplied to the public elsewhere lags well behind American levels, as does the growth of new electronic media.[33] The displacement of print words by electronic ones that occurred in the United States in the 1970s may still lie in the future for other advanced countries.

The declining dominance of print media is a cause for concern, for they are the media that in the United States and elsewhere in the free world enjoy autonomy from government. It matters that people are increasingly getting their news and ideas through governmentally controlled media. The fact that media enjoying the full protection of the First Amendment, or of its equivalent traditions abroad, have ceased to dominate the information marketplace is in itself troublesome.

Yet if the print media remained totally free and unaffected by what is happening to the electronic media around them, there might be little cause to worry, for the realm of print, though quantitatively reduced, would remain a realm of freedom. Perhaps the continued existence of one forum for uninhibited debate would be enough to assure a ferment of opinions and ideas. If so, it would suffice that free speech in the traditional media continued unaffected by the regulations applied to the new media. That, however, is not the case. The new media are not only competing with the old media for attention, but are also changing the very system under which the old media operate.

____ 3. Electronics Takes Command

Once upon a time companies that published newspapers, magazines, and books did very little else; their involvement with other media was slight. They reviewed plays and movies, they utilized telephones and telegraphs, they reported on the electrical industry, but before about 1920 they had limited business ties with any of those industries. This situation is changing, and with implications adverse to freedom.

A process called the "convergence of modes" is blurring the lines between media, even between point-to-point communications, such as the post, telephone, and telegraph, and mass communications, such as the press, radio, and television. A single physical means—be it wires, cables, or airwaves—may carry services that in the past were provided in separate ways. Conversely, a service that was provided in the past by any one medium—be it broadcasting, the press, or telephony—can now be provided in several different physical ways. So the one-to-one relationship that used to exist between a medium and its use is eroding. That is what is meant by the convergence of modes.

The telephone network, which was once used almost entirely for person-to-person conversation, now transmits data among computers, distributes printed matter via facsimile machines, and carries sports and weather bulletins on recorded messages. A news story that used to be distributed through newsprint and in no other way nowadays may also be broadcast on television or radio, put out on a telecommunication line for printing by a teletype or for display on the screen of a cathode ray tube (CRT), and placed in an electronic morgue for later retrieval.

Technology-driven convergence of modes is reinforced by the economic process of cross-ownership. The growth of conglomerates which participate in many businesses at once means that newspapers, magazine publishers, and book publishers increasingly own or are owned by companies that also operate in other fields. Both convergence and cross-ownership blur the boundaries which once existed between companies publishing in the print domain that is protected by the First Amendment and companies involved in businesses that are regulated by government. Today, the same company may find itself operating in both fields. The dikes that in the past held government back from exerting control on the print media are thus broken down.

The Electronic Revolution

The force behind the convergence of modes is an electronic revolution as profound as that of printing. For untold millennia humans, unlike any other animal, could talk. Then for four thousand years or so their uniqueness was not only that they could move air to express themselves to those immediately around them but also that they could embody speech in writing, to be preserved over time and transported over space.[1] With Gutenberg a third era began, in which written texts could be disseminated in multiple copies. In the last stage of that era phonographs and photographs made it possible to circulate sound and pictures, as well as text, in multiple copies. Now a fourth era has been ushered in by an innovation of at least as much historical significance as the mass production of print and other media. Pulses of electromagnetic energy embody and convey messages that up to now have been sent by sound, pictures, and text. All media are becoming electronic.

The word "electronic" implies more than electrical. The dictionary tells us that electronics is the science of the behavior of electrons in gases, vacuums, or semiconductors. Long before the vacuum tube was invented, telegraph and telephone messages flowed over electrical wires, but the flow was not subject to much control. Except for noise and attenuation, the current that went in was the current that came out. From Lee de Forest's vacuum tube of 1906, through computers, to today's computers on a chip, progress in electronics has allowed manipulations such as storing, amplifying, and transforming electrical signals. In today's electronic communication, the electrical pulses are stored in computer memories as arrays of on-off switches, which are usually represented in verbal explanations as zeros or ones and called bits. Transmission from sender to receiver also flows in such digital codes. To say that all media are becoming electronic does not deny that paper and ink or film may also continue to be used and may even be sometimes physically carried. What it does mean is that in every medium—be it electrical, like the telephone and broadcasting, or historically nonelectrical, like printing—both the manipulation of symbols in computers and the transmission of those symbols electrically are being used at crucial stages in the process of production and distribution.

Each revolutionary technology—writing, printing, and electricity—was clumsy and limited in its early stages. Writing on cuneiform tablets could not compete with alphabetic writing on paper, nor could Gutenberg's press compete with a rotary one. The seeds of the electronic revolution lay in the late eighteenth century when scientists found that electric currents could travel far. One of the first uses of this phenomenon that occurred to them was for signaling. A variety of telegraphs was invented. In 1774 in Geneva, George-Louis Le Sage strung a circuit for each letter between two rooms, in one of which a switch could be opened or closed on each circuit, thus agitating in the other room a pith ball labeled with a letter. By watching the sequence of jiggling balls, the receiver could read an alphabetic message. In 1807–1808, Samuel T. von Soemmering built a similar device in Munich with water jars instead of pith balls. Whenever a circuit was closed in one room, electrolysis started in one of the letter-labeled jars in the other room, causing bubbles to rise.[2]

To Samuel Morse goes credit not for having the telegraphic idea, which was widespread, but for launching in 1844 an economic form of telegraph which actually caught on. His competitors Royal E. House and David Hughes, like Le Sage, had made alphabetic systems that would mechanically reproduce each letter; but with the technology of the day, such equipment was clumsy and prone to break down. Morse adopted a simple key to make and break a circuit, producing dots and dashes of sound depending on the operator's timing. The rugged device required highly trained operators, but labor in his day was cheap. What would be the wrong solution for today was the right solution for then.

Even with Morse's simple device, early telegraphy was expensive. One message at a time traveled over hundreds of miles of electrical wires strung on poles across the countryside. At a high cost per word, only the most valuable messages were telegraphed, and they were transmitted in the most cryptic form possible. Code was used to abbreviate common phrases into words. In newspapers the front page often had a short column of telegraph "bulletins," each about an inch long. Lengthier analytic coverage waited for stories sent by the more economical mails. The practical uses for telegraphy in its early form were therefore limited.

The cost of telegraphy could be cut if several messages could be sent at once at different tones over the same wire. This was called a multiple telegraph. Venture capitalists funded Alexander Graham Bell to invent such a device. In the process he developed the telephone, using varied tones to do more than carry a few simultaneous telegraph signals.

The early telephone, too, was a crude and limited device. To keep costs down, a fidelity level was accepted well below that at which music could be enjoyed or the phone used for entertainment. The system was optimized for its single most marketable use, conversation.

Broadcasting became economically possible only half a century later. Music and entertainment of tolerable quality were delivered at costs people could afford by using the free airwaves instead of expensive wires and switches. Such open broadcasting was not adapted to private person-to-person communication, and even communication for small audiences was impracticable for early radio and television. A few stations were all that the system then allowed, and these were allocated to uses of high priority. Such uses included point-to-point communication for ships at sea or for the armed forces. For the general public in a democratic country another high priority was broadcasting for mass audiences.

The technological constraints on electronic communication are now becoming less confining. Alternative transmission systems such as the telephone network, cable systems, and microwaves can all be used in a great variety of ways. They can be used at low cost with low fidelity if that is sufficient to the need, or at higher cost for higher fidelity. They can be used point-to-point or broadcast. They can be encrypted for privacy or transmitted in the clear. The design of communications systems today need not follow a cookbook recipe for each purpose to be served but is more a matter of optimizing among the several alternatives for which a multipurpose system may be used in a mix of markets. Rarely will the optimum arrangement be a wholly separate system for each purpose. More often, different uses and groups of users will share many of the same facilities.

For the first three-quarters of the twentieth century the major means of communications were neatly partitioned from each other, both by technology and by use. Phones were used for conversation, print for mass distribution of text, movies for dramatic entertainment, television for entertainment too, radio for news and music, and phonograph records for music. No law of nature said that it had to be that way. Other approaches were proposed and tried. For a quarter of a century from 1893 phones carried music and news bulletins to homes in Budapest, Hungary. Thomas Edison thought that the main use for the phonograph he had invented would be for mailing records as letters. Drums and bugles have been used not just for music but to send messages to distant auditors. But there were good reasons for the success of one device for one use and another device for another use. Cost, hability, and the existence of better alternatives all determined what a device would be used for.

The fact that different technologies were consecrated to different uses protected media enterprises from competition from firms using other technologies. Newspaper publishers did not worry about the growth of phonograph records; each kept to its own turf. The separation of modes was never complete, but it created significant moats.

Now the picture is changing. Many of the neat separations between different media no longer hold. IBM and AT&T, which once thought themselves giants of different industries, now compete. Each can provide customers with the means for sending, storing, organizing, and manipulating messages in text or voice. Cable television systems no longer just distribute broadcast programs but also transmit data among business offices and sell alarm services, movies, news, and educational courses. Enterprises that in the past saw themselves as in quite different businesses now find themselves in competition.

The explanation for the current convergence between historically separated modes of communication lies in the hability of digital electronics. Conversation, theater, news, and text are all increasingly delivered electronically. Electronic methods have proved superior not just for exotic new uses but also for communications that in the past were done by the physical impact of ink on paper. Sound and images can be sampled and transmitted as digital pulses. Using computer logic on such arrays of bits, large complicated patterns representing text, voice, or pictures can be manipulated by computer with far more flexibility than was possible with paper or earlier electrical but analog records. Such digital records can be preserved in electronic memories, converted in format, and transmitted instantaneously to remote destinations. Thus all sorts of communications processes that in the past were handled in unique and cumbersome nonelectronic ways may now be mimicked in digital code. All sorts of communications can therefore be carried on the same electronic network.

Depending on circumstances, either intensification of competition or monopoly among carriers may follow from this fact. With the fences down, battles are fought for what were once separate turfs. Whether an enlarged monopoly emerges or competition remains the norm is determined partly by policy, but also very largely by the characteristics of the technology. The emerging technology is likely to foster only a few small islands of monopoly, since many alternatives exist for the delivery of every service.

There is nothing new about attempts to use communications technologies for a variety of purposes. The savings in doing so are obvious. The problems were technical, and it is these problems that digital electronics is overcoming. It is thereby bridging three main technical separations that existed between established industries— between the telegraph and telephone, between the telephone and radio, and between print and electrical delivery. Through these mergers electronic technology is bringing all modes of communications into one grand system.

Convergence of Telegraph and Telephone

When the telephone was invented in 1876, it immediately affected telegraphy and, later, the postal system too. The telegraph companies were at that point trying to move beyond a Morse code system for sending dots and dashes from operator to operator to a teleprinter system that would provide a switched network for sending written messages from the premises of the sender to the premises of the receiver, as telex does today. Inventors were trying to improve on the House and Hughes printing telegraphs to make them practical, and overcome the situation in which, to send a telegram, one had to go in person to a telegraph office where a professional telegrapher encoded the message. Business premises frequently had electrical buzzers hooked up to the neighborhood telegraph office to call a messenger to pick up outgoing messages, but for the ordinary citizen there was no such convenience. As early as 1848 *Punch* magazine argued that the telegraph was "too good a thing to be confined to public use" and that it should be introduced into "the domestic circle." Ten years later *Punch* returned to the theme of a telegraph with lines reaching to the customer's home, or at least "within a hundred yards of every man's door." "With a house telegraph," *Punch* concluded, "it would be a perpetual tête-à-tête."[3]

There was a recognized need for a person-to-person discourse device. By the mid-1870s entrepreneurs were beginning to introduce printing telegraphs into business offices. In Great Britain the ABC Telegraph Company was doing that. In the United States Western Union extended the quotation ticker of the Gold and Stock printing telegraph service into brokerage and related offices. The introduction of telephone service aborted these developments. The market for end-to-end telegraphs collapsed, and the development of a switched printing telegraph system in America was delayed for half a century. Clumsy, mechanical nineteenth century typewriter terminals could not compete with the little black phone in person-to-person service.

From the invention of the telephone on, the history of the telegraph and telephone systems has been one of continuous competition and convergence.[4] In 1876 when Western Union declined to purchase Bell's telephone patent for $100,000, Western Union was the largest corporation in the United States.[5] By 1910, the now larger telephone company turned around and bought control of Western Union for $30 million. From the beginning, the developers of the Bell System conceived of a "grand system" that would include both voice and text communication.[6] Bell's 1876 patent was written to cover a device that could provide both telephone and multiple telegraph service over a wire line. The original charter of AT&T, written in 1885, was designed to cover a communications system, not just a telephone system, and the name chosen for the company included both Ts.

There were technical reasons for combining the telephone and telegraph systems. Each required a wire plant. There were tens of thousands of miles of telegraph lines already in place. The iron wire that had been used proved not fully adequate for telephony, but for limited service over short distances it could serve; and the poles could be used.[7] Copper telephone lines could carry telegraph code as well as voice, and at the same time.

In 1910, the year of the merger, President Theodore Vail of AT&T described the new arrangement as "One system with a common policy, common purpose and common action; comprehensive, universal, interdependent, intercommunicating like the highway system of the country, extending from every door to every other door, affording electrical communication of every kind, from every one at every place to every one at every other place." But this hope was not to succeed. The era of trust busting had begun, and the federal government brought proceedings. AT&T was forced in 1913 to divest Western Union.

The next AT&T effort at using the telephone plant for text transmission was the introduction in 1931 of a switched teletype service, called TWX. Teletypes of acceptable reliability were already being used in the 1920s on specially leased lines by press services and financial institutions. But when AT&T offered a public switched teletypewriter service, once again the Department of Justice intervened and forced the sale of the TWX system to Western Union, to be merged with its similar telex service.[8]

But phoenix-like, the issue of text over telephone lines will not die. Its present form concerns communication among computers. In the United States computers may be connected to each other and to terminals either by dedicated lines leased from a carrier such as AT&T or Western Union, or by the public dial-up switched telephone network which everyone uses. Computer networks now transmit far more text than does telex. In America, with some millions of computer terminals and only about 100,000 telex terminals, transmissions over computer networks in 1977 carried more than 10,000 times as many words as did telex. So now, after a century during which many fingers wagged "no," the telephone system has in fact become what its creators always thought it would be, a hybrid telephonic and telegraphic system, carrying both human voice and written messages over the same lines.

Convergence of Telephone and Radio

Over-the-air transmission and enclosed transmission can converge in either or both directions. Signals like phone calls that used to be carried on wires might move onto microwave radio channels, or signals like television broadcasts that used to be sent over the airwaves might be put on cables, or indeed a mix is possible. Whether for a particular stretch it is preferable to use a microwave or an enclosed carrier is now just an economic engineering choice. Exactly the same things can be done either way. In the past, various unsuccessful efforts were made by practitioners of wire or radio technology to use their own medium to compete in activities adapted to the other; now it can be done. In the past the limitations of the technology controlled the use; now the use controls the technology.

It seemed at one time that just as telephone had undercut the telegraph, so radio might become a threat to telephony. Radio transmissions spanned distance without requiring an investment in wires. So for AT&T, radio was both an opportunity and a competitive threat. If radio telephony turned out to be practicable, and if it came to be controlled by rival entrepreneurs, the Bell System's "grand design" for a "universal service" would become a fleeting dream.

On Christmas Eve in 1906, R. A. Fessenden demonstrated the possibility of sending voice by radio to ships at sea. AT&T's Vail and his chief engineer, John J. Carty, recognized a looming danger. It appeared that just as wire telegraphy had spawned wire telephony, what was called radio telegraphy might spawn radio telephony. Yet they also suspected that wireless telephony would not be a satisfactory substitute for the existing wired system. In 1907 Vail assured a London banker that "the difficulties of the wireless telegraph are as nothing compared with the difficulties in the way of the wireless telephone." Still, in 1909 Carty asked for funds for research on a telephone repeater that "might put us in a position of control with regard to the art of wireless telephony, should it turn out to be a factor of importance." In the end, it proved true that over-the-air radio was at the time incapable of supporting a national point-to-point transmission network, but to the telephone company's leaders it had been a close call.

One result was adoption by AT&T of an aggressive research and development policy designed to cover the entire field of electrical communication and, more particularly, to occupy the "shoulders of the field" at the edges of telephony. Carty organized a new research division with a heavy focus on wireless technology. In 1910 the company employed 192 engineers in development work, with a budget of just under $½ million; by 1916 it employed 959 men, with a budget of over $1½ million. In 1925 the Bell Telephone Laboratories were formed specifically to do fundamental research in communication on the borderlands of conventional telephony. Out of Bell Labs came much of the progress in radar and the transistor as well as in such nontelephonic modes of communication as sound movies and television.[9]

Despite its technical leadership, the telephone system was not able to deliver entertainment. This was not for lack of trying. There were dozens of preradio attempts to provide what Asa Briggs has called "the pleasure telephone."[10] Most of these efforts at offering something like modern radio or Muzak over wires never got beyond the idea stage or, if they did, went bankrupt in short order.

In retrospect, it is easy to see why their economics was faulty. The fidelity of tone needed for entertainment would have greatly increased the cost of ordinary phone service. More important, in 1896 in New York City basic telephone service cost about $20 a month; the income of an average worker at that time was about $38.50 a month. Clearly, few people were yet in the market to become subscribers. But to deliver service over wires economically, as every cablecaster knows, it is essential to get subscriptions from a large proportion of the households that a wire passes. A cable system or other wired entertainment service to 5000 or 10,000 customers might be profitable if they all are located cheek by jowl along a single line, but very unprofitable if they are scattered all over a city and each has to be reached by a separate line.

Radio broadcasting eliminated this problem. The broadcaster could start without a mammoth investment. The few early radio enthusiasts might be scattered around a city, but any and all of them could be reached at a fixed and modest transmitter cost. Over-the-air transmission was clearly the system of choice in the early days of entertainment broadcasting.

When high levels of penetration are reached, however, the balance of costs between the airwaves and wires is more even. In the 1920s, for example, planners in the Soviet Union found it cheaper to install wired loudspeakers than to market radios.[11] In the systems of independent radio receivers and wired loudspeakers, the common element is the loudspeaker. In the radio system one has to add a tuner, amplifier, and antenna. To the speaker in the wired system, one has to add some yards of wire. Which one is cheaper clearly depends on the number of yards of wire. In the Soviet planned economy the number of yards per speaker could be small. If the authorities decided to have an apartment house wired for speakers, or to put speakers down a certain street, they could ensure a high proportion of subscribers along that line. So for forty years the wired speaker was the dominant device for radio entertainment in the Soviet Union. Not till 1964 did the number of regular radios in the country come to exceed the number of wired loudspeakers.

Decades later, development in China followed the same course. The wired loudspeaker in the 1950s and 1960s was a major feature of the Chinese landscape, but in the 1970s the transistor radio spread.

The invention of the transistor swung the balance strongly in favor of over-the-air radio broadcasting. Similar developments occurred in noncommunist countries. Redistribution systems, as the wired systems are there called, are fairly common in European cities. Before the transistor they were common for radio loudspeakers, and they are now for television.

The idea of television over wires is an old one. As early as 1912 a sociologist of science, S. C. Gilfillan, published an article on "The Future Home Theater" outlining the two alternative ways of bringing television to the home, by radio or by telephone wires.[12] By the beginning of television broadcasting in the 1930s the economic choice was clear. For audio signals, ordinary cheap wires are an adequate transmission medium. For television transmission, the capital requirements are greater. A single television channel can indeed be transmitted over ordinary electric wires, called "twisted pairs," but if viewers are to have a choice of channels, they need either several such pairs serving them, or a remote switch that they can operate to choose what is to come over their wire, or else service by a carrier with greater bandwidth than wires, such as the coaxial cables that were invented in the 1930s.[13] The word "bandwidth" describes the range of frequencies, or bits per second, that a carrier can transmit. The single-toned dots and dashes of a telegraph can be handled on an iron wire. The larger range of frequencies required to reproduce the human voice needs a better wire. To reproduce a moving picture requires several hundred times as much bandwidth as does speech if there is to be satisfactory picture definition. Many video channels at once can travel on the thick carrier called a coaxial cable, or on the beams of the very short radio waves called microwaves, or on the still shorter waves of light that can be guided through a glass fiber.

The attraction of cable television is not that it is cheaper than over-the-air television even at high penetration rates, but that it is cheaper for a higher grade of service. It is the cheapest way of getting either a clearer picture or many more channels to each subscriber. The issue in the marketplace is whether the public wants that enhanced service enough to pay for it. In some places the answer clearly is yes. In Belgium and Canada, penetration is very high because cable enables viewers to see television shows from across the border. In Belgium, the attraction is programs from France for the French-speaking population. In Canada, where the majority of sets are on cable, the motive for subscribing is that cable allows people who are at a distance from the frontier to receive American network programs. In the United States, about 30 percent of television homes presently receive their service by cable, and the number grows a couple of points or more each year. The main motive to subscribe was initially to improve the picture in locations where reception was poor or to receive all three networks in places where all did not broadcast. But then a new attraction was offered in the form of sports, movies, and other programs on pay channels. Over thirty-five syndicated services are today delivered to American cable systems by satellite. New cable systems are being built with over one hundred channels so as to accommodate the growing variety of programing.

Cable has thus become a viable system for broadcast delivery. Where it exists, it offers far more bandwidth than can be made available over the air. Not all of that bandwidth need be used for entertainment. Already in a few places like lower Manhattan the cable system is offering banks and businesses data transmission service on its lines. Once cable system penetration gets so high that the lines are everywhere, cable can become an alternative to the telephone system's local loop to the customer for all uses for which the switching of circuits at the telephone exchange is not crucial. In sum, until the development of repeaters and coaxial cables that were cheap enough and good enough to maintain the quality of signal along the cable, the delivery of video over an enclosed medium was not an economically attractive option. Now that it is, the shift of some delivery from over-the-air to enclosed media has begun.

The movement in the other direction, from wire lines to over-the-air radio, which so worried AT&T in 1907, took a step forward during and just after World War I. The American government feared that Great Britain, which formed the hub of the world undersea cable network, might gain similar leadership in wireless.[14] So it backed the formation of a strong radio manufacturing company, the Radio Corporation of America (RCA), with the help of and with access to the patents of General Electric and AT&T. But when radio turned out to be useful for mass entertainment broadcasting, AT&T set out to gain control of that new service for itself.

So a struggle for control came to a head between RCA and AT&T. The phone company saw radio transmission as a natural extension of its communications business and sought to extend its monopoly into that area too. In the tradition of a common carrier, AT&T had no intention of getting into programing. It developed a plan whereby it would build a transmitter and studios in every city and, for a fee, make them available to whoever wished to broadcast.[15] But AT&T had not thought through where the customers who desired to buy air time would come from. As an experiment, in 1921 the phone company established station WEAF in New York. If the experiment was commercially successful, stations would be established elsewhere and networked. It turned out, however, that because radio receivers were still few, there was little incentive for anyone to buy and program air time. As a result, broadcasters themselves had to take responsibility for putting on programs. That was RCA's plan for creating a broadcasting system, and so it won the battle.

The possibility of networking became the key to a settlement between AT&T and RCA. In 1926 AT&T abandoned broadcasting in return for RCA's commitment to use AT&T's lines rather than Western Union's for networking . The agreement provided for exchange of patent licenses, and WEAF was sold to RCA, where it became WNBC, the flagship station for the National Broadcasting Company.

Dramatic evidence of networking's importance came in 1923 when, on the death of President Warren Harding, Calvin Coolidge's eulogy was carried by long-distance telephone lines to broadcasting stations around the nation and thus to an audience of millions. Network radio, which reached coast to coast in 1925, ultimately led to a growth in presidential power through Franklin Roosevelt's fireside chats. The philosophy of localism in broadcasting, which underlay the license allocation plan of both the Radio Act of 1927 and the Communications Act of 1934, was somewhat vitiated by the development of broadcasting networks, a system in which both the broadcasting companies and the telephone company played a part.

Thus from 1926 until communication satellites came along in the 1970s, telephone and radio companies carved out separate but complementary domains, with the radio companies doing the programing and transmission to listeners and viewers, while AT&T provided the national networking circuits via cables and microwaves. But this divorce may not last. Just as after a century of separation, computer communication finally forced the marriage of text and voice, so too by a century after 1907, broadcasting and telephony may merge.

One step toward that merger is the development of cable television. For the moment the broadcasts are distributed on a different network from the one used for conversation, but convergence may occur in a variety of ways, some allowing broadcasting to be done over the phone system, and others allowing telephony over cable systems, as well as over the airwaves. All these are now possible. Radio techniques that did not exist in 1907 allow thousands of simultaneous point-to-point voice or text transmissions to take place over the air without interference.

One of the new techniques that have changed the picture is the use of microwaves. The medium-length waves used for ordinary broadcasting on the AM band, which were in use before the 1930s, tend to follow the surface of the earth and so interfere with each other at long distances. But ultra-high-frequency waves (UHF) and the even shorter microwaves travel in straight lines, like light, and so do not pass beyond the horizon. Interference in their case is local, and hence the same frequencies can be reused beyond the horizon.[16] Furthermore, radio waves can be focused in a beam so that they will not spread out in all directions. The size of the antenna required to create such a beam is a function of the length of the waves. Antennas used to direct short waves may require four 400-foot towers spread over a quarter of a mile. The antenna to direct the much shorter UHF waves may be only 60 feet long atop a tall tower. The antenna to beam the still shorter microwaves may be just a portable dish 3 to 12 feet across. This makes it practicably cheap to send microwaves in narrowly focused beams. Beams pointed in different directions will not interfere with each other.[17]

Microwaves today carry most long-distance phone traffic. They are beamed from tower to tower or to and from satellites. Microwave circuits may also come to carry much local phone service, at least wherever substantial bandwidth is needed, being beamed from rooftop to rooftop. A system like this, with local nodes where the traffic is concentrated, is called digital termination service (DTS). It is employed by large telecommunications users as a way to start and end their long-distance calls by establishing microwave links between their offices or plants and the long-distance toll switch. It may entirely bypass the local phone company.[18]

Another advance in utilization of the airways is use of very low-power transmitters for local carriage. A weak radio-telephone signal can be sent from a subscriber to a nearby concentrator which multiplexes (or transmits simultaneously) many signals on a broadband circuit. The new cellular systems for radio telephones that serve moving vehicles are examples of such use of low-power, very local transmission.

Land-mobile radio communication with cars, taxis, and trucks has been growing rapidly, to the point where congestion in large cities is severe. The cellular solution to this problem is to place low-powered transmitters in each cell of a checkerboard grid. The cells may have a radius of miles or of blocks. Each of the cell transmitters is connected to a central exchange having a computer. As a subscribing car drives from cell to cell, it sends out an identifying radio signal that is received at the cell's transmitter node, so the computer always knows where the car is. A phone call for the car is routed to it via the transmitter in the cell where the car is currently located. The frequencies used by these weak transmitters can be repeated every few cells without interference, allowing radio communication to many vehicles simultaneously without congestion. So far, the only use for which cellular radio has been licensed is mobile communication, but in principle the same idea could provide over-the-air competition with the local telephone loop, though not yet at a competitive cost.

Other technical developments that would allow thousands of users of the airwaves at once in point-to-point communication are by sophisticated forms of multiplexing. The earliest method used for putting several messages on a circuit at once was frequency division multiplexing. Graham Bell had used it to send messages over the same wire at the same time, by having each set of dots and dashes sound a different note. Frequency division is also the way broadcasters are separated. Each has a share of the spectrum centered on a different frequency. If broadcasters are silent at any moment, their frequency range is wasted. Frequency division multiplexing is therefore not an efficient system. More complicated systems allow many users to share a frequency.[19]

Time division multiplexing of various kinds allows multiple users to share the same frequency, but with each using different time slices of it, which may be only thousandths of a second long. Time division multiplexing is attractive for digital communications, because computers can keep track of this fantastically precise timing. Time division multiplexing is just as useful for communication over coaxial cables and optical fibers as for communication over the air and thus may keep these enclosed transmissions economically competitive with radio.

Spread spectrum techniques allow a message to be sent using a wide range of frequencies which are simultaneously used by other users. At some frequencies, interference occurs and the message gets blocked. At other frequencies, it gets through and the receiving set recognizes the messages addressed to it. Until recently this technique was used only by the military for reliable and secure communication in conflicts, but it is beginning to have civil applications.

These three techniques—beamed microwaves, low-power transmitters, and sophisticated multiplexing—all permit a great increase in the number of users who may simultaneously be on the air without interfering with each other. They allow radio to be used for much of what the terrestrial wire network was used for in the past, while avoiding its cost. At the same time the enclosed terrestrial carriers are becoming so much more efficient that they may be used for much of the traffic, such as broadcasting, that in the past was economic only over the air.

Cable systems are just the first step. Future phone systems may also compete with cable systems and broadcasting to deliver broadband services to the home and office. By the use of optical fibers, coaxial cables, compression techniques for coding the same information in fewer bits, or even just by enhancement of the copper wire pair, the local loop may end up handling a much higher bandwidth than is available to the phone subscriber today. Entertainment programs, news, and education may thus be delivered to the home via a multiplexed enclosed channel, and broadcasting may be removed from the air onto that network. The over-the-air spectrum may be reserved for such uses as reaching moving vehicles or satellites to which no trailing wire can be attached.

Some forecasters have suggested that the convergence of telecommunications and broadcasting will take the form of one integrated digital network serving all purposes. But large-scale communications activities will always justify specialized facilities that are optimized for their particular use. Airplanes and air traffic controllers will talk to each other over different equipment from what is used to carry movies to homes. Satellites will continue to carry long-distance messages, while switched wires, coaxial cables, or optical fibers will give local service. Broadband facilities will be used for multiplexed trunk lines, computer-to-computer communications, or video, while those who need only narrow-band signals may seek cheaper facilities. But if these diverse facilities are all electronic, they can all be interconnected. Narrow-band signals can be multiplexed onto broadband trunks for part of their trip, perhaps ending up on a cheap local line. Like a road system, the communications network will have its superhighways, feeder roads, and private driveways. But they will all make an interconnected system. One way or another, the competition among technologies that Vail saw on the horizon in 1907 has been slowly moving closer and will soon be here.

Convergence of Print and Electronics

In the past, a broad moat separated the print media from the electrical ones. These two technologies did things differently enough to limit competition between them, though to a degree they competed in news, fiction, and pictures. The competition and convergence of electrical media with print began with telegraphy, continued with broadcasting, and today is most striking with data or computer networks.

From the infancy of radio, newspapers wondered how broadcasting would affect them. With Lowell Thomas and H. V. Kaltenborn proclaiming the news to millions, it was easy to speculate that this new and easy way of keeping informed might kill or at least injure newspapers. An alternative hypothesis held that a short newscast would stimulate interest and lead listeners to read newspapers more. Research demonstrated that the second proposition was closer to the truth.[20] There are some people in the world who are interested in news and others who are not. Most of the people who listened to newscasts, as well as those who read the newspapers, were the ones who were interested. Radio and the printed page were not substitutes for each other but opportunities for the news-hungry, who used them both. And insofar as following news built a habit, the media supported each other. The same sort of relationship existed between radio sports and attending games, records and attending concerts, radio drama and taking related books out of the library. Indeed, it became somewhat of a cliché among communication researchers to say that different media did not displace but reinforced each other.

Then came television. The motion picture industry, though it survived, was fundamentally transformed. Attendance plummeted. And when a football game was shown on the air, the stands were often empty. Clearly the relationship among media is more complicated than always giving mutual support or always displacing one another. The question now is which pattern applies to newspapers, magazines, and books in confrontation with computer information services. Is the relationship one of reinforcement, like that between newspaper and radio news, or is it one of cataclysmic reconstruction, like that between movies and television?

A major factor determining whether the mechanism of support or displacement predominates is the degree to which the new medium fully serves those needs that its rival medium satisfied. Hearing a radio report of a ball game is a lesser experience than watching it. But watching it on television may be every bit as exciting as being there, perhaps more so. The functions that a medium serves are not just its ostensible ones but also latent ones, frequently unrecognized. A pioneer British study, for example, found television to be a substitute for movies more for young children than for adolescents; for adolescents, going to the movies was a chance to leave home and go out on a date.[21] Nor is the interaction between media encompassed merely by the complementary or substitutive reactions of the audience. For example, in the United States, television combined with rising postal rates killed such general magazines as *Life*, not because people would not consume both, but because both media sought support from the same advertisers, and the advertising pie was not large enough to feed both.

For the past century there has thus been continual ebb and flow among media as changes in price, technology, and character have interacted on one another. The development of photography in the nineteenth century gave magazines a giant boost. Newspapers responded with telephotographs and rotogravure sections. Those sections were then killed by movie newsreels. The newsreel in turn could not survive television. Newspapers used to carry serialized fiction, which in many countries they no longer do because magazines and television outdo them in that function.

But while television beats the print media in some domains, it also sustains the print media. The broadcasting industry has its trade journals. Newspapers run broadcasting schedules, thereby winning readers. *TV Guide* has the biggest weekly circulation of any magazine in the United States, and *Radio Times* and *TV Times* have the largest magazine circulations in England. In the pretelevision era, movie magazines were among the most successful ones.

Media react on each other not only in their rise and decline but also in change of character. In an era in which broadcasters get the news out fastest, newspapers have had to give their readers features and analysis that the broadcasters do not provide.[22] Newspapers now rush less to bring out the news of the minute; they no longer issue extras. But they do much more of what weeklies also do in filling in background and examining the news in depth.

Though there has been both convergence of and competition between electronic and nonelectronic media for some time, now something is new. It appears likely that digital electronic networks may, in the twenty-first century, carry the bulk of what is today delivered as printed paper. The output may still often end up as words on paper, but the paper is likely to be spewed out at a terminal to which the information has flowed electronically. The old totally separated system of print publishing, in which hard copy is produced by the mechanical pressing of ink on paper and the copy is delivered by physical carriage, is being challenged by electronic technologies.

Today, as yesterday, mail is still for the most part carried physically by vehicles and men. Printed publications too are physically moved, sometimes by the post. But such physical portage of text is rapidly giving way to electronic transmission. Already within companies a large part of the message traffic comes off computer terminals. There are few functions for which the classical technology of print remains unconnected to electronic transmission, or will long remain so. The costs at almost every stage of electronic publishing and record keeping are becoming lower than those for hard copy. The time is fast approaching when handling information on paper rather than by computer will be an extra cost alternative done only for taste or convenience.

Not too far in the future hardly anything will be published in print that is not typed on a word processor or typeset by use of a computer. The long and bitter struggle of compositors' unions to protect linotyping is over. Periodicals and publishing houses compose their text on computer terminals and edit it that way too.

The electronic transformation of printing has been most dramatic in newspapers. For a century, news services have sent text to the editorial room electrically, but in the past the electrical representation was lost as fast as hard copy sheets rolled out of the teletypewriter. Now the news service feeds not only a printing terminal but also a computer's memory where the text is retained for editing. On many papers reporters no longer type their stories on paper at all. They type at a cathode ray tube (CRT) and edit right on the screen, leaving the story in the computer for further processing. The editor reviews the stories at his CRT. The formating, headlining, placement, and composition of ads are also done on computer terminals that have been developed specially for the newspaper industry. Ironically, the first paperless offices in the world, if there are any, may be in newspapers.

One result is that the entire newspaper exists in computer readable form before it ever rolls off the presses in print. The computer tape need not be and probably is not thrown away. It has many uses. For one thing, newspapers can now be indexed by programs that search the tape for key words. This is useful in the market for information services. The publisher can sell to a researcher a compilation of all the stories that have dealt with some specified topic. So newspapers can move into the information retrieval business. The first outstanding example was the *New York Times* Information Bank.

Soon everything that gets printed will exist also in computer form, and filing will be more efficient in the electronic than the manual form. With electronic publishing there are also writings that in the Gutenberg sense never get printed at all. Texts entered on one processor can be read on a screen at that or another linked processor without ever being printed. These are only the early steps in the electronic transformation of publishing.

In the coming era, the industries of print and the industries of telecommunication will no longer be kept apart by a fundamental difference in their technologies. The economic and regulatory problems of the electronic media will thus become the problems of the print media too. No longer can electronic communications be viewed as a special circumscribed case of a monopolistic and regulated communications medium which poses no danger to liberty because there still remains a large realm of unlimited freedom of expression in the print media. The issues that concern telecommunications are now becoming issues for all communications as they all become forms of electronic processing and transmission.

Cross-Ownership

The technological convergence of modes is the major force bringing publishing into the regulated electronic environment. A contributing factor is the development of conglomerates which are in the business of both print and electronic communications. These have been largely separate businesses in most democratic countries, sometimes because of laws and regulations restricting cross-ownership between communication modes. In most countries the electronic media are government monopolies, but the print media are private. The post, telephone, and telegraph are usually in one ministry, though more often than not the "P" and "T" parts of the ministry hardly talk to each other. Broadcasting is often a separate government monopoly, and where commercial broadcasting exists, sometimes there may be restrictions on cross-ownership between publishing and broadcasting.

The partitions to the communications industry are more numerous in the United States than elsewhere. An elaborate attempt has been made by the FCC, the Department of Justice, and the courts to encourage pluralism by partitioning communications. There are half a dozen main partitions that they have tried to preserve against convergent trends in the technology, but with limited success. More often than not, the barriers have not lasted. Technology frequently makes the separations inefficient, and entrepreneurs seek to expand their empires' sectors.

One separation that the American government long tried to maintain was between telegraphy and the competing media of telephony and the mails. Since 1913 the federal government has sought to preserve Western Union against entry of the Bell System into telegraphy. But with the decline of telegraphy and the growing cost of messengers, Western Union has been forced to rely on the postal service and AT&T for its delivery system. Mailgrams, which by 1977 exceeded telegrams in words sent by a factor of five, go by teletype to a post office near their destination for delivery from there by the postal service. Ordinary telegrams are now mostly phoned. The sender of a telegram from Boston to New York dials a Boston phone number but is connected to a clerk in a Western Union office in New Jersey. The clerk transcribes the text and phones it to the New York addressee. The system is an anachronism.

With the arrival of computer data transmission, the advantage of using the phone network for computer traffic became so obvious, and vested interests in computer use so strong, that the government no longer tried to exclude the phone company from that business. The traditional separation was broken. In prospect is an ever larger role of the phone network in text transmission. Today numerous companies compete to provide long-distance service for both voice and text. These include not only AT&T but also Satellite Business Systems, International Telephone and Telegraph, and MCI Communications Corporation. The number continues to grow.

Another separation that the government tried long to maintain was between domestic and international carriers. That border too has now been breached. In 1960 Western Union was compelled to divest itself of Western Union International and was given a monopoly in domestic telegraphy by the FCC; a set of companies called international record carriers—specifically WUI, RCA, ITT, and TRT—were given by the FCC the international cable business to themselves. Both the FCC and Congress have changed their minds about the wisdom of that arrangement, and Congress has changed the law. The Record Carrier Competition Act of 1981 allows Western Union and the international record carriers to compete both domestically and internationally, and they are doing so.

Until the 1980s AT&T, which did handle both domestic and international voice traffic, was not permitted to use the telephone lines for international data traffic. However, technology in the form of acoustic couplers, which enable users to connect any computer terminal to an ordinary telephone, made that regulation unenforceable. The FCC has abandoned it, so AT&T is now both domestically and internationally in both the voice and data communications business.

A domestic-international division has also been maintained in satellite communications. The United States is committed by treaty to using Intelsat, the international satellite organization, for international satellite transmission. For domestic satellite traffic, however, the FCC has adopted an "open skies" policy. Within the United States, a competing set of companies have been licensed to provide communication by satellite; these include RCA, Western Union, the American Satellite Corporation, Satellite Business Systems, Hughes, and Comsat General.

Technically, the segregation between international and domestic satellite transmission is artificial. The perimeter of a satellite's beam does not follow national boundaries, and a beam can cover as much as one-third of the earth. There may be political reasons but there are no engineering reasons to bar anyone from using a satellite to communicate anywhere within its beam. So the distinction between domestic and international satellites has not been consistently maintained. For example, Intelsat now provides domestic satellite service for at least seventeen countries. Algeria was the first. Its towns deep in the Sahara had no telephone service to the cities along the Mediterranean shore. By putting a ground station in each of these towns and also near the capital and contracting with Intelsat for circuits, Algeria instantly established communication without the expense of constructing a line of microwave towers through the desert. Conversely, Canada's Anik 1 domestic satellite provides international communication. Canada led the United States by a year in having a domestic communication satellite, during which time some American customers used Anik for communication between points in the United States. They have done so since in periods of transponder shortage on American domestic satellites.

Various proposals are currently being made that would further blur the distinction between domestic and international satellites. Satellite Business Systems, an American specialized common carrier with a satellite of its own, is negotiating with Canada to offer service there too. Indonesia's Palapa satellite is being used also by the Philippines, Malaysia, and Thailand. Regional satellite systems are being proposed for the Middle East, the Andean bloc, East Asia, and other groups of nations. Both the limit to the number of satellites that can be placed on the geostationary orbit under present frequency allocations and the cost of satellites make shared use attractive for small or medium-sized countries. And in the United States the philosophy of deregulation makes the FCC more willing to allow a process whereby domestic and international carriers can go into competition with each other.

Another partition that regulators have tried unsuccessfully to make stick is that between communications and computing. What made that distinction impossible was the evolution in the 1960s of time sharing, in which many different users at scattered locations all use the same computer at once. Such computing systems are also telecommunication systems. Almost every time-sharing system has a mailbox program that allows persons on the system to send messages to each other. Under the Communications Act of 1934, telecommunications carriers have to be licensed by the FCC, but the FCC was neither empowered nor eager to turn the computer industry into a regulated industry or to take upon itself the task of regulating it. The FCC therefore had to find a way to distinguish time-sharing systems from communications networks. In 1971, in Computer Inquiry I, the FCC announced official definitions distinguishing computing from communications. But the distinctions failed. Every digital communication system that stores and forwards messages does things that fall squarely under the definition of computing, and conversely, every time-sharing system with its mailboxes and access from a distance does things that fall squarely under the definition of communications.

So in 1976 the FCC made a second try. In Computer Inquiry II, the dichotomy between computing and communication was dropped. Instead, a distinction was drawn between "basic" and "enhanced" communications services, which is also an uncertain distinction. New inquiries have been started to try to draw lines in the borderland between them.

In the meantime, companies in the business of communications and computing are each invading the terrain of the other. IBM is a partner in Satellite Business Systems. Other computer manufacturers, such as Control Data Corporation, have formed data networks to service their customers. Telecommunications companies, particularly those that introduce new digital switches, which are themselves computers, may offer their customers computing services on their facilities. This is the kind of "enhanced" service that AT&T will increasingly offer now that the 1982 consent decree and Computer Inquiry II have freed it from the 1958 prohibition against engaging in any business but common carrier communications. The divested local operating companies will still be prevented by the new consent decree from marketing the powerful computing capabilities of their digital switches for other computer services, but for how long?

Cross-Ownership of Publishing and Broadcasting

Of all the government efforts to police cross-ownership, those most actively promoted and debated have involved broadcasting stations or cable systems. Ever since the 1920s, one consideration weighed in granting broadcasting licenses has been the other businesses in which the applicant was engaged, and in particular, whether to give broadcasting licenses to newspapers.

One aim of restrictions on radio-press combinations is to maintain a plurality of voices. The goal is to foster organs of expression for varied ideologies and points of view. But it is hard to establish separations by point of view without getting deeply into evaluation of the content. For regulators who do not wish to become arbiters of content, it seems easier to require separations based upon the technology used. That, however, reminds one of the story of the drunkard's search. A drunk is under a street light looking for a lost key. A passerby asks, "Did you drop it here?" "No," the drunk says, "I dropped it over there, but the light is better here." Separations by technology are easier to enforce than separations by ideology, but whether they also serve the goal of maintaining a plurality of points of view is dubious.

In all developed countries the number of newspapers has been falling. The number of cities in the United States with genuinely competing newspapers declined to thirty by 1981. A few other cities had two papers owned by the same chain, but in most there was simply one paper. In 1923, 39 percent of the cities had two or more papers; ten years later, the figure fell to 17 percent; in 1943, to 10 percent; in 1953, to 6 percent; in 1963, to 4 percent; and in 1973, to 2 percent. In 1910, only 62 American newspapers belonged to chains; the rest were independently owned. By 1980, 1139 of the 1745 dailies belonged to chains.[23]

But do not jump to conclusions! There is a great deal of competition left in publishing, particularly in books and magazines, but even in newspapers; in many respects it is not declining much, and in some ways even growing. The average size of a chain in 1910 was 4.7 newspapers; the average size of a chain in 1980 was 7.4 newspapers.[24] So there are still a large number of separate publishers, most of whom are rather small. In 1923 the largest 1 percent of newspaper firms had 22.6 percent of the circulation. In 1978 the largest 1 percent of newspaper publishing firms had only 19.8 percent of the circulation.[25] Newspapers from nearby towns compete in the territory between them, and metropolitan and suburban papers compete in the suburbs.[26] Still, the bulk of the American population lives in cities with only one newspaper having a significant circulation.

The really important competition for newspapers is with other news media: the weekly news magazines, the few serious national newspapers such as the *Wall Street Journal* and the *Christian Science Monitor*, radio, and television. Against these increasingly powerful competitors the press is seeking to protect itself. The major tactic is to buy into the competition. If you can't lick them, join them. The Gannett chain of local newspapers has launched a national paper, and the *New York Times* prints editions around the country. Major newspaper chains buy magazines and broadcasting stations. Both Dow Jones, which publishes the *Wall Street Journal*, and the *New York Times* have entered the computer data base business too. The Knight-Ridder newspaper chain has tested out videotex in Coral Gables, Florida, providing information by phone lines onto television screens.

There is nothing new about such tactics. In the 1920s when news-

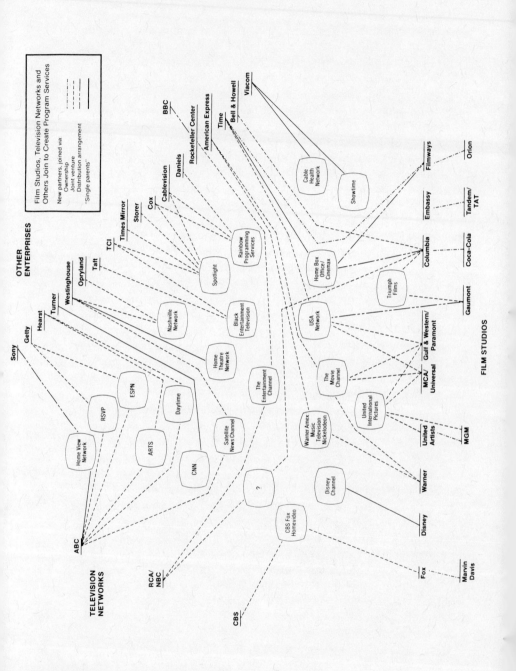

OTHER ENTERPRISES

Film Studios, Television Networks and Others Join to Create Program Services

New partners, joined via
Ownership
Joint venture
Distribution arrangement
"Single parents"

TELEVISION NETWORKS

FILM STUDIOS

papers started worrying about the fledgling news medium of radio, publishers turned to the strategy of acquiring broadcast licenses so as to ride both horses. They could make a strong case that no one was better equipped to provide a good news service or to cover community activities. Indeed, many of the better stations have been newspaper owned. In 1923, 12 percent of the country's 576 stations were publisher owned. The percentage rose to 29 percent by 1945. Since then, the trend has been mixed. Radio stations owned by newspapers dropped to 7 percent of the much increased total by 1979, but 30 percent of the television stations were still newspaper owned, though down from a peak of 42 percent in 1950.[27]

Although the trend is not one-way toward concentration, cross-ownership is clearly significant (see figure). In consequence a large number of organizations, though free of government regulation as publishers, find themselves also producing heavily regulated and licensed media, and therefore looking over their shoulders at their regulators. Time, Inc., owns not only magazines but also a television station, ATC (the largest multisystem cable operator), USA (an advertiser-based cable service), and Home Box Office (the largest pay-cable service). The Times-Mirror Company earns only 47 percent of its income from the *Los Angeles Times* and its other five papers. It also owns seven television stations and the seventh largest multisystem cable operator. The Gannett newspaper chain owns seven television stations, Scripps-Howard six, and McGraw-Hill five. Conversely CBS, ordinarily thought of as a television network, owns ten magazines with a total circulation of over thirteen million.[28] The current proliferation of mixed press-broadcast-cable joint ventures occurs despite the fact that public policy restricts newspaper-broadcaster cross-ownership.[29]

The FCC and the courts have found it to be against the public interest to allow any one company a *de facto* newspaper monopoly along with a major voice in broadcasting in a particular town. A company may own a paper in one town and a broadcasting station in another, but since 1975 it may no longer acquire both in the same town. Existing cross-ownership arrangements are generally grandfathered until the station is sold, after which the rule applies.[30] To a degree, government policy has worked. According to one estimate, the number of different voices available from separately owned newspapers, radio, and television stations increased by 25 percent between 1950 and 1970.[31] FCC policy, however, is limited to increasing separate voices within single communities. There is no bar to the same company owning different media in different cities.[32]

Under the First Amendment no government agency may regulate what business activities a newspaper or other publisher engages in or make the absence of cross-ownership a condition of publishing. The way in which the FCC can make rules about cross-ownership is through its issuance of broadcast licenses. Through its handle on broadcasting it achieves a regulation of print that it could not achieve directly. It would be hard to argue, under the constitutional imperative of free speech, that the advocate of some cause could be prohibited from publishing a message in several media at once, such as in a pamphlet, on an audio cassette, and on a videotape. But on the grounds that there can be only a few broadcasting stations, the case has been made for not allocating broadcast licenses to individuals who already own other media in the same place.

Since the 1920s the most lively cross-ownership issue for the press has therefore been its involvement with broadcasting. Radio and television were the media that threatened the press and regarding which it sought to hedge its bets. Now on the horizon is a new set of threats arising from entirely new media. Information retrieval systems, such as the on-line business news services found in brokers' offices or the videotex being experimented with in many parts of the world, are new competition for the press. The publishers are worried. One group of newspapers in Texas recently went to court to stop AT&T from running a videotex experiment. Newspaper publishers have been trying to persuade Congress, and have succeeded in persuading the judge in the Bell System antitrust case, to prohibit AT&T from publishing informative kinds of electronic yellow pages.

The first defensive tactic by the owners of an old medium against competition by a new one is to have the new one prohibited. If this does not work, the next defensive tactic is to buy into the attacker. Newspapers tried to limit radio news. In Great Britain, at the newspapers' behest, the British Broadcasting Corporation (BBC) was forbidden for years to do anything with news except read a short wire-service bulletin: no detail, no drama, no effects, just a drab announcement. But in the United States, where the tactic of prohibition was not available, cross-ownership was tried. Print publishers became media conglomerates. Broadcasters for years tried to stop cable television; now they want to be cablecasters too.

Cross-Ownership of Cablecasting and Other Media

In cable television cross-ownership is evolving rapidly. American cable enthusiasts generally wanted to keep that medium out of the hands of broadcasters and phone companies, two competitors with motives to stifle its challenge to their established way of doing things. In 1970 the FCC banned phone companies from owning cable systems in their geographic area of operations and banned television networks from owning cable systems anywhere. In 1971 the FCC also banned television stations from owning cable systems within their reception area. The networks are now fighting those rules, and telephone companies may be expected to fight them in the future. In 1981 the FCC granted CBS a modest exception, allowing it to buy cable systems for up to 90,000 homes for experimentation. In 1982 the FCC issued a notice of proposed rulemaking to eliminate the prohibition on network ownership of cable systems.

A still more important issue about cross-ownership is whether the owner of a physical cable may be the programer of the system or must be only a common carrier. The consequences of allowing a cable system to be owned by a company from some other medium of communications depend critically on whether cablecasting is a business of carriage only or of content too. On the one hand, the physical cable along the streets is likely to be a monopoly. If the company which has this monopoly also controls the programing, there is reason for concern, and even more so if the company also controls other media. On the other hand, there is no inherent limit on the number of channels of cable programing. One company's having a channel does not prevent another company from having a competing channel. Policy considerations regarding cross-ownership of the physical plant and the channels for use are thus quite different.

In the 1980s the issue of how to organize a mature cable system is likely to be a major national controversy. Cable is now organized as a broadcasting system rather than as a common carrier. Much more money is to be made by selling movies and sports on pay television than by leasing channels for others to program. Cablecasters therefore prefer to see themselves less in the communications engineering business than in the entertainment business. Investors in cable systems are largely companies in mass media rather than in electronics. In 1981, 38 percent of cable systems were owned by broadcasting companies, a quarter or more by publishing organizations, 3 percent by theater owners, and another 3 percent by phone companies.[33]

While the notion of having the physical cable run by a common carrier is popular among thoughtful observers of cable in the United States, the assumption is widespread that it would be a carrier other than the telephone company. The expectation in Great Britain has been the reverse, though the issue is now being debated. The time will come when broadband digital phone systems can provide the same service as a cable system. When this happens, there will no longer be any need for two feeds into each home, one for voice and one for video. The telephone company will become an obvious carrier for video as well as voice. The issue of cross-ownership restrictions on telephony and cable will then become acute.

When the time comes, whether with optical fibers or with cable, that hundreds of channels of video can reach each home, there will no longer be any justification for banning cross-ownership of individual channels or of programing so long as the number of channels is kept above the number normally in demand for lease at reasonable cost. If a newspaper wishes to improve its service by coupling its print offering with an on-line data base, or with a newsreel, one could only cheer this result. Certainly the preservation of the press, a goal that everyone proclaims, is better served by encouraging the print media to make imaginative use of multimedia opportunities than by enjoining them from doing anything but putting ink on paper in an obsolete conventional way.

The death of American railroads occurred partly because they saw themselves as being in the business of moving trains over steel rails rather than of moving goods by whatever technology was most efficient. Newspapers can face the same sort of fate. Imaginative newspapers today know this and are seeking to be in the information business regardless of the channel. Their health may easily depend on being in the dissemination business via a cable carrier as well as via print on paper, and via other sorts of delivery media too. This is one reason that the issue of whether cable systems have common carrier obligations or are instead monopoly broadcasters or publishers is so important, and will be so fiercely fought between the conflicting information industries in the near future.

Implications for a Free Press

The regulations on cross-ownership of electronic media are striking testimony to the extraordinary pervasiveness of government regulation. Some of the segmentations seem sensible, while others seem absurdly wrong-headed. Some are designed to diversify sources of information in the community, while others serve no purpose but to protect an established vested interest. But useful or not, these requirements are in principle at odds with the First Amendment tradition of no government authority. They would be totally unconstitutional as applied to print. Yet over the electronic media, government has exercised its authority in ponderous detail.

More often than not, however, the bars on cross-ownership have proved ineffective. Despite government attempts to prevent it, cross-ownership reinforces a convergence among modes that is drastically changing the structure and legal status of communications industries. Neither competition among modes nor convergence between them is new; what is new is the scope of the convergence. Today, thanks to science's mastery of the mechanics of transmission of electronic and optical energy, engineers are able to transform and retransform the frequencies and amplitudes of signals at will. They can convert signals between analog and digital mode to utilize the advantages of each; they can convert to whatever frequencies are most convenient. Communications of all kinds, when transformed into digitized form, can be not only transmitted but also stored and modified as desired. Computers can manipulate communications in all media so as to synthesize graphic patterns or voice, to edit text or video tapes, to write abstracts without a human author, to draw inferences, or to find proofs. Thus a broadband digital communication system is likely in the end to become a vehicle not only for data communication but also for other kinds of communication, including voice and pictures, publishing, broadcasting, and mail.

It is misleading to ask where all this will lead, for there is no steady state to which the process of unification and differentiation will lead. The trend as far ahead as anyone can see is toward a convergence among media, with great communications institutions working in many modes at once in interconnected ways. Convergence does not mean ultimate stability, or unity. It operates as a constant force for unification but always in dynamic tension with change. New devices will be invented to serve specialized needs. There will always be specialization, innovation, and attempts to do differently and for some purposes better what a universal telecommunication system does, and there will always also be a return to the universal system because of the extraordinary convenience of universality.

There is no immutable law of growing convergence; the process of change is more complicated than that. Nonetheless, a particular trend of convergence has been set in motion by the development of electronic communication. Before that, a series of institutions existed based on the printing press, namely newspapers, magazines, and books. Over the centuries a legal tradition developed that protected these from government control. Electronic communication has not been equally well protected by its newer tradition. The extension of electronic means to do better and faster what the older modes of communication did with lead, ink, and paper has inadvertent consequences for the sustenance of freedom.

4. The First Amendment and Print Media

Communication under the First Amendment was to be an activity unregulated by government. "Congress," the Amendment says, "shall make no law . . . abridging the freedom of speech or of the press." Since the adoption of the Fourteenth Amendment, the states too have been barred from depriving any citizen "of life, liberty, or property without due process of law." This formula, the courts have held, extends the application of the First Amendment from being a restriction on the Congress only to one on the states as well.[1] These two articles of the Constitution thus create for the American people a realm of "uninhibited, robust and wide-open" discussion unfettered by public regulation.[2]

The First Amendment's simple bar to legislation, in twenty-five words or less, has engendered millions of words of interpretation. One school of thought, the absolutist view of Justices Hugo Black and William Douglas, takes the First Amendment at face value. "No law," said Justice Black, means "no law."[3] In numerous bristling dissents these two judges argued that if Congress may make no law abridging speech, it may make no law against pornography, obscenity, libel, or sedition. That absolutist view is not just an idiosyncratic quirk of a couple of recent judges. Thomas Jefferson maintained: "Libels, falsehood, and defamation, equally with heresy and false religion, are withheld from the cognizance of federal tribunals."[4] And James Madison described the amendment as a "positive" and "absolute reservation."[5]

The absolutist view, however, is not the court-enforced law of the land; and the issue here is not how the First Amendment ought to be interpreted, but how it has been interpreted in the context of different technologies. The dominant interpretation of free speech and a free press has at all times been more convoluted than the rendering of the absolutists. Despite the ritual praise given to free speech in civic rhetoric, the libertarian tenet of the First Amendment has never in the real world of politics had national consensus behind it, and certainly not at the beginning. In various opinion polls, only a minority of respondents support free speech for those whose views are anathema.[6] A favorite experiment for students in social science courses is to take radically libertarian quotes from the founding fathers and mislabel their source. Most respondents reject the quotations when not attributed to a respected author. In every generation attempts have been made to impose censorship or to ban unpopular views.

Commentators have repeatedly distinguished liberty from license and interpreted the amendment as protecting only the former, not the latter. There are always those, at times including the majority on the Supreme Court, who cannot conceive that the intent of the amendment is to deny to the government the power to prevent speech that seems patently vicious, harmful, or dangerous. The conservative Federalist Joseph Story said with extraordinary bluntness that the First Amendment guarantees only that "every man shall be at liberty to publish what is true, with good motives, and for justifiable ends."[7] Few governments, if any, have claimed a need to interfere with that!

But the Supreme Court rejected this eviscerated view of the meaning of the First Amendment even more decisively than it has drawn short of the absolutist view. The historical meaning it has given to the amendment through its interpretations is a complex but strong one, incorporating a ban on prior restraint, strict procedural requirements for any regulation of speech, and a presumption against such restriction.

Prior Restraint

In American legal jargon the words "prior restraint" refer to what laymen are more likely to call censorship or licensing. The phrase refers to any government action to review, permit, or prohibit a publication before the act of publication has taken place. In the nineteenth century a common restrictive interpretation of the First Amendment confined the clause to being just a ban on prior restraint. Justice Isaac Parker of Massachusetts, for example, said in 1826 that the freedom constitutionally guaranteed was designed merely "to prevent all such previous restraints upon publication as had been practiced by other governments" and not to "prevent the subsequent punishment of such as may be deemed contrary to the public welfare."[8] English legal history and Blackstone's *Commentaries* were heavily relied on to limit the First Amendment to nothing more than the requirement that prosecution must follow on evil speech and not anticipate it. This restrictive interpretation of the First Amendment disregarded the face meaning of its words and took them instead in a historical context.

Justice Felix Frankfurter was the most recent learned expositor of a restrictive historicist view of the amendment. In 1951, he quoted as the "authentic" view of the Bill of Rights a judgment from 1897: "The first ten amendments to the Constitution, commonly known as the Bill of Rights, were not intended to lay down any novel principles of government, but simply to embody certain guarantees and immunities which we had inherited from our English ancestors, and which had from time immemorial been subject to certain well-recognized exceptions . . . In incorporating these principles into the fundamental law there was no intention of disregarding the exceptions, which continued to be recognized."[9] Libel, profanity, obscenity, and sedition were among the "exceptions" Frankfurter had in mind when he used this quotation to sustain the conviction of leaders of the American Communist Party for organizing to advocate the overthrow of the United States government.[10] Conversely, among the inherited "guarantees and immunities" which he and the entire Court recognized as having been won by "our English ancestors" was the absence of prior censorship on publication.

The prohibition on prior restraint was developed to its logical limit in Justice Charles Evans Hughes's 1930 judgment in Near v. Minnesota.[11] "The liberty of the press," he said, quoting Blackstone, "is indeed essential to the nature of a free state, but this consists in laying no *previous* restraints upon publications, and not in freedom from censure for criminal matters when published."[12]

The defendant in this case, J. M. Near, had been publishing a paper, *The Saturday Press*, which was as good a target for suppression as there could be. Its main thesis in every issue was that the chief law enforcement officials in Minnesota were in cahoots with Jewish gangsters, all named.[13] The case for libel was patent. But the Minnesota law under which Near was prosecuted was not the libel law; it was a law which provided for an injunction to be issued against malicious or scandalous newspapers. The Minnesota judges from whom the injunction was sought had no difficulty finding *The Saturday Press* malicious and scandalous, and they issued the injunction.

A minority of the Supreme Court agreed with them. Justice Pierce Butler argued that freedom of the press did not include the right to perpetrate libel, that Near had had his day in court in the hearings on the injunction, that what he was doing was a violation of the freedom of others, and that on the basis of the evidence the injunction could properly be issued. He maintained, in words that sound much like the complaints of some recent Presidents and Vice-Presidents, that the Minnesota judges had a right to be concerned about the "terrors of the press, introducing despotism in its worst forms." A majority of the Court, however, joined with Hughes in finding the Minnesota law the epitome of what was not allowed under the Constitution, because it provided not for the punishment of past libels but for the suppression of future issues of a paper as yet unseen.

A variety of prior restraints on speech and publication have been struck down by the Court. In 1937, for example, the Court invalidated a municipal ordinance requiring a permit before distributing literature door-to-door.[14] In 1945 the Court rejected a state law requiring registration by union agents entering the state to organize.[15] In 1958 the Court outlawed an ordinance requiring dues-charging organizations to get a permit before soliciting.[16] As Justice William J. Brennan warned in 1963, "Any system of prior restraints of expression comes to this Court bearing a heavy presumption against its constitutional validity."[17]

More recently, in 1971, the Court refused, on grounds that it would be prior restraint, to issue a permanent injunction on publication of the Pentagon Papers.[18] It did, however, issue a temporary injunction while hearing the case and allowed for the possibility of injunctive action in extreme circumstances of threat to national security, which in the situation before it the Court failed to find.[19] In similar fashion, a lower court escaped a decisive ruling on an extreme case in 1980 when *The Progressive* magazine planned to publish information on how to build an H-bomb. Once again a temporary injunction was issued while the case for a permanent injunction was being heard. But before the judge could reach a decision, the same information was published elsewhere, so that no secret was left to be protected, and the injunction was dropped.[20] How seriously the courts take prior restraint on public affairs publications is evidenced by the fact that it took something as horrendous as the dissemination of secrets of nuclear weapons construction to cause the courts seriously to consider allowing precensorship.

While an important part of the meaning of the First Amendment is the ban on prior restraint, the Supreme Court has rejected the argument that this is all there is to freedom of speech or of the press. The Court majority, which has repeatedly held that freedom of speech is not absolute, has nevertheless narrowly defined the limited verbal acts which might by legislation be held to be abuses, even after the fact.

Since there is some speech that the government may not regulate, even after the fact, and other speech that it may, some test is needed to distinguish between the two. Four such tests have been used at different times by the Court. One of these tests is that of clear and present danger. A second test balances freedom of speech against other legitimate public interests. A third test distinguishes between protected and unprotected types of speech, such as between discussion of public affairs, which the First Amendment is said to protect, and emotional expression or commercial use of words, to which it is said by some not to apply. Recently the Court has to a degree lessened its reliance on these three tests. The fourth test, which it still unanimously accepts, distinguishes between words that are simply speech *per se* and those that are an integral part of a regulatable action.

Clear and Present Danger

A major step in defining the limits of constitutional interference with speech was the formulation in 1918 of the "clear and present danger" doctrine by Justice Oliver Wendell Holmes in Schenck v. United States.[21] As has happened so often in the history of the Court, the decision was given to one side; the reasoning that became a precedent was given to the other. Charles T. Schenck, the general secretary of the Socialist Party, who had been convicted for issuing pamphlets against the draft, lost his case, but in the decision Holmes laid out a logic by which that finding was an extreme exception, and by which in ordinary times the defendant would have been within his constitutional rights.

Holmes argued that the character of every act depends upon its circumstances: to shout fire in a theater cannot be construed as merely speech. "The question in every case is whether the words used are used in such circumstances and are of such a nature as to create a clear and present danger." Schenck's propagandizing against the draft in wartime, in the opinion of the Court, fell afoul of that test. More important than the validity of the Court's evaluation of what Schenck did was the fact that by implication the Court denied to Congress under the First Amendment the right to legislate about speech where no clear and present danger exists. The Court's wording in this case was nevertheless cautious: "It well may be that the prohibition of laws abridging the form of speech is not confined to previous restraints, although to prevent them may have been the main purpose." Thus a Supreme Court precedent for protection of speech even against subsequent punishment had been set, except under the limited circumstances of a clear and present danger.

With the passing of years the Court has fluctuated in defining the scope of those limited circumstances that it is willing to recognize as meeting that test. The formula of clear and present danger has a Januslike character: it can be used as a defense of free speech against legislative attempts to curtail unpopular views, but it can equally be used, whenever judges become sufficiently alarmed, to justify an exception to First Amendment immunities. The formula came to be accepted by all wings of the Court, but with very different interpretations by the so-called conservatives (notably Frankfurter) and the First Amendment absolutists (notably Black and Douglas).

In several cases during the late 1940s and 1950s the Court sustained the perception of Congress that the Communist movement was not just a group of people with unconventional views but rather a disciplined conspiracy operated from abroad which thus represented a genuine enough danger to justify regulatory measures, such as requiring registration by Communist organizations and the filing of non-Communist affidavits by union officers as a condition for enjoying the advantages of the National Labor Relations Act.[22] Following Schenck, those who issued antidraft literature in wartime were held to represent a clear and present danger.[23] And in situations where police feared a riot as a result of provocative speeches, the Court on occasion found interference or controls to be justified.[24]

At the same time, the Court has declared that the danger must be very real to justify restriction of speech. As Justice Black argued in 1941: "What finally emerges from the 'clear and present danger' cases is a working principle that the substantive evil must be extremely serious and the degree of imminence extremely high before utterances can be punished . . . It must be taken as a command of the broadest scope that explicit language, read in the context of a liberty-loving society, will allow."[25]

The Januslike character of the clear and present danger test is visible in the increasingly vehement dissents of Frankfurter, on the one hand, and Black and Douglas, on the other, as they perceived the majority (in their opinion) misusing the doctrine. Frankfurter, in a series of historical disquisitions, sought to rescue Justice Holmes from what he thought to be misinterpretation which the Court was giving his decisions: " 'Clear and present danger,' " he said in 1946, "was never used by Mr. Justice Holmes to express a technical legal doctrine or to convey a formula for adjudicating cases. It was a literary phrase not to be distorted by being taken from its context."[26] Black and Douglas, on the other hand, finally reached the point in 1969 of rejecting the application of the clear and present danger test.[27] If Congress could make "no law" abridging freedom of speech and of the press, this meant "no law" and did not mean that it could make such a law when faced with a clear and present danger.

The Court has not accepted this absolutist position any more than it has Frankfurter's position. It continues to use the phrase "clear and present danger" but uses it less.[28] Recently, in situations where in the past it would have referred to speech as representing a clear and present danger, it has tended to note the relationship of speech to some illegal action.

Balancing Public Interests

The disagreement between the First Amendment absolutists and the "conservatives" on the Court centered not only on the clear and present danger test but also on the inclination of conservatives such as Frankfurter to weigh off other public concerns against that represented by free speech. Frankfurter condemned "those who find in the Constitution a wholly unfettered right of expression." Congress and the courts, he argued, must in each situation balance the injunction of the First Amendment against other legitimate national concerns, such as security, order, and repression of obscenity: "The demands of free speech in a democratic society as well as the interest in national security are better served by candid and informed weighing of the competing interests."[29]

This notion of balancing was considered by the First Amendment absolutists to be in flagrant disregard of the unequivocal language of the First Amendment.[30] According to Justice Black, "the First Amendment does not speak equivocally. It prohibits any law abridging freedom of the press."[31] In 1959, he wrote: "To apply the

Court's balancing test . . . is to read the First Amendment to say 'Congress shall pass no law abridging freedom of speech, press, assembly and petition, unless Congress and the Supreme Court reach the joint conclusion that on balance the interest of the Government in stifling these freedoms is greater than the interest of the people in having them exercised.' "[32] While libertarians may regret that there were not nine Justice Blacks on the Court, there were not; neither his straightforward reading of the words of the First Amendment nor Frankfurter's pragmatic reinterpretation of those words into vacuity became the Court-enforced law of the land. Although the Court majority has repeatedly held that freedom of speech is not absolute and has sustained strictly defined laws against obscenity and libel, restrictions on pretrial publicity, and controls on Communist infiltration, still the Court has given to First Amendment considerations a special status, treating them not merely as considerations to be balanced off evenly against others.[33]

In a variety of respects the Court gives precedence to First Amendment considerations in making the balance. At a conceptual level, this weighting is expressed by the Court's assertion that freedom of speech enjoys a "preferred position" in the law of the land.[34] Operationally, this preferred position means that for those who claim interference with their First Amendment rights, certain procedural burdens are waived and certain usual legal presumptions are reversed.

The Court has followed at least nine different rules in giving a preferred position to First Amendment rights:[35]

Reduced presumption of constitutionality
Shift in the burden of proof
Expedited actions
Disallowance of vagueness
Requirement of well-defined standards
Disallowance of overbreadth
Disallowance of procedural burdens
Restriction on choice of means
Narrow interpretation of laws

Rule one is that in First Amendment cases the presumption of constitutionality of government action is reduced. Normally the Court starts with a presumption that a legislative action is constitutional. If a case can be resolved without testing the constitutionality of a law, the Court will avoid such a test. If there is a reasonable argument that the law serves a purpose within the legislature's power, the court will not gainsay it. If the text of a law can be made constitutional by choosing one alternative interpretation of it, then that interpretation will be chosen.

When ordinances and laws are challenged on First Amendment grounds, however, the Court has on occasion invalidated them as facially unconstitutional, without examining whether, in the particular facts of the case, the legislation could have been sustained under some interpretation of the law's wording. The Court has not allowed a law to remain in force if its application would in some circumstances oppress free speech. For example, an ordinance barring nudity in motion pictures shown in drive-in theaters, if visible from the street, was held unconstitutional because the ordinance did not explicitly specify that the ban applied only to obscene nudity. The Court would not supply that interpretation.[36] Where First Amendment rights are at stake, the burden is on the legislators to draft the rules carefully; the Court will not interpret them generously.

Rule two is that in actions against speech the burden of proof is firmly on the state.[37] A California statute making a bookstore owner responsible for stocking an obscene book, regardless of his awareness, was held unconstitutional, though a food store owner could by law be made responsible for the quality of foodstuffs on the shelves. Where the result may restrict free expression, the Court places the burden of proof on the prosecution.[38] In a remarkable ruling in 1982 the Supreme Court allowed students in the Island Trees Union Free School District to challenge in court the exclusion of nine controversial books from the school library. Though the school board normally has authority to choose the books it buys, it was forced to bear the burden of proving that it was not seeking to deny the students "access to ideas."[39]

A third rule in First Amendment cases is not necessarily to await final action by administrative bodies and lower courts before the higher courts act. Ordinarily the Supreme Court does not consider a case until all other legal remedies are exhausted and final action has been taken in the courts below. In First Amendment cases, however, the Court may step in early, for delay in allowing people their rights is itself a denial of rights. The ability of the government to delay a publication can be a serious abridgment of the freedom of the press.

No longer need defendants risk violating the law and then take their chances appealing a conviction, for the Court now sometimes intervenes in an ongoing case to halt a proceeding which, by the act of prosecution alone, whatever the outcome, will have a chilling effect on the exercise of freedom.[40]

Vagueness is another vice in legislation that the Court does not tolerate when it creates a threat to First Amendment rights.[41] Legislation vaguely conveying authority for government action, which would pass muster if the challenge were only on due process grounds, may be overturned if the challenge is on free speech grounds.[42] One form of vagueness is the failure to offer well-defined legislative standards for executive regulation of speech. For movie censorship, parade permits, or similar regulations of speech to stand, the authorizing statute must specify criteria for refusal in clear language. The applicants must be able to know their rights and obligations, and the administrators must be clearly guided to confine their action to constitutional goals and methods that ensure due process. Their discretion must be strictly limited.[43]

Another form of vagueness that has fatally faulted various ordinances and statutes is to state an acceptable prohibition so broadly that it also covers activities exempted by the First Amendment along with ones that the legislature may properly prohibit. It is within the police power of the state to prohibit public use of fighting words that create a danger of breach of the peace, but simply to prohibit public use of fighting words is too broad. Those words may sometimes be used in situations where there is no danger. An ordinance potentially impinging on First Amendment rights must be explicit in limiting its application to situations where state action is allowable.[44]

Procedural burdens that would ordinarily be acceptable may become intolerable in the eyes of the Court if they impinge upon free speech. For example, formalities that result in extensive delays have sometimes been found to be unconstitutional violations of free speech.[45]

If there are alternate means to achieve a legitimate legislative goal, a legislature may not pick a means that restricts free speech. Since the Court majority admits of circumstances in which rival public concerns make the abridgment of free speech legitimate, it finds itself having to examine the reasonableness of legislative enactments restricting speech. One rule used by the Court when a legislature, seeking a proper goal, has chosen a means that impinges on speech is to inquire whether other means are available to achieve the same end. If there are, the Court may reject the choice of means that raises constitutional difficulties.

Arkansas, for example, passed a law requiring school teachers to fill out a form listing all organizations to which they belonged, defending the law on the grounds of the state's interest in employing teachers of high character who would be good influences on the children. The Court rejected this means toward a legitimate end. Justice Potter Stewart argued: "In a series of decisions this Court has held that, even though the governmental purpose be legitimate and substantial, that purpose cannot be pursued by means that broadly stifle fundamental personal liberties when the end can be more narrowly achieved."[46]

Justice Owen Roberts formulated the same principle in 1939: "Where legislative abridgment of the rights is asserted, the courts should be astute to examine the effect of the challenged legislation. Mere legislative preferences ... may well support regulation directed at other personal activities, but be insufficient to justify such as diminishes the exercise of rights so vital to the maintenance of democratic institutions."[47] Ordinarily, if the legislative goal is a proper one, the Court does not question the lawmakers' judgment about the means chosen to achieve it. When the chosen means threaten free speech or free press, however, the Court may bar their use.

Finally, the Court maintains the preferred position of First Amendment rights by choosing narrow interpretations of statutes or ordinances to keep them within First Amendment bounds. This is the reverse of the tactic by which the Court denies the usual presumption of a law's constitutionality. Here it tries to save a law by truncating its unconstitutional applications. The Court is not confined to one tactic; it has precedents for proceeding in either way.[48]

In these nine rules the Court has affirmed its concern about the constant impulse of those in authority to undermine free discussion. Whereas Black and Douglas spoke in dissent more often than not on First Amendment matters, the majority joined with Black in saying that the First Amendment "must be taken as a command of the broadest scope that explicit language, read in the context of a liberty-loving society, will allow."[49] And whereas the Court did not follow Black in refusing to balance other considerations against free speech, it did, when seeking such a balance, apply a heavy weight in favor of free speech wherever "the case confronts us . . . with the duty . . . to say where the individual's freedom ends and the State's power begins. Choice on that border, now as always delicate, is perhaps more so where the usual presumption supporting legislation is balanced by the preferred place given in our scheme to the great, the indispensable democratic freedoms secured by the First Amendment."[50]

Protected and Unprotected Speech

A strategy used by the Court majority to justify some government regulations on speech has been to distinguish certain kinds of speech, for which the First Amendment is said to be intended, from other kinds for which it is not. The absolutists would apply free speech to any kind of expression, but the majority view sustained, for example, a law forbidding public name-calling and abusive language: "There are certain well defined and narrowly limited classes of speech, the prevention and punishment of which have never been thought to raise any constitutional problem. These include the lewd and obscene, the profane, the libelous, and the insulting or 'fighting' words—those which by their very utterance inflict injury or tend to incite to an immediate breach of the peace."[51]

The Court also upheld an Illinois statute creating an offense of libel against races or religions, not just against individuals. Frankfurter wrote the decision with hardly any reference to the First Amendment (as Black and Douglas protested), for he saw the issue as one of libel law and therefore unrelated to protected speech, which in his view did not include libelous accusations.[52] Similarly, in a decision on obscenity, Justice Brennan in 1957 argued that the purpose of the First Amendment is to protect free discussion of "matters of public concern." Obscenity *per se* is emotive and not a statement of views. Citing the existence of laws against blasphemy, profanity, libel, and obscenity in the founders' times, he argued that "the unconditional phrasing of the First Amendment was not intended to protect every utterance." The clear and present danger test applies, he said, as the limit of restraint to speech that is protected by the First Amendment, but obscenity is not protected speech at all: "All ideas having even the slightest redeeming social importance—unorthodox ideas, controversial ideas, even ideas hateful to the prevailing climate of opinion—have the full protection of the guarantees, unless excludable because they encroach upon the limited area of more important interests. But implicit in the history of the First Amendment is the rejection of obscenity as utterly without redeeming social importance."[53] This dictum led pornographers to embed smut in messages with some "redeeming social importance" and thus acquire constitutional protection.[54] So the decision, while doctrinally antilibertarian, in fact opened the doors to pornography.

The Court, finding that the criterion of "redeeming social importance" was met by virtually every publication, retreated to a sterner set of standards. In doing so, it persisted in attempting to define an unprotected realm of pornography which is not speech under the First Amendment.[55] The Burger Court of today is more puritanical than its immediate predecessors.[56] A Court minority, however, recognizes the impossibility of separating statements having a message from those whose only purpose is titillation. "The Court," Justice John Marshall Harlan wrote, "seems to assume that 'obscenity' is a peculiar genus of 'speech and press' which is as distinct, recognizable and classifiable as poison ivy." Because the Court has been so divided in recent cases on obscenity, and because of the more liberal perspectives on sex in the generations now growing to maturity, the view that there is no objective basis for the Court to distinguish obscene from other expressions seems bound to prevail—if not the stronger Douglas view that all speech, obscene or not, is protected.

In such policy debates, the rhetoric by which advocates justify their positions is most often either historical or philosophical. Both approaches have been used to delimit speech that is protected by the First Amendment from classes of speech which are not. Older legal reasoning relied heavily on precedent, turning to eighteenth century practice to find discriminations. Forms of speech which the colonies and states then regulated, such as libel, profanity, blasphemy, sacrilege, or sedition, were, it was reasoned, not intended to be placed above the law by the First Amendment.

During the mid-decades of the twentieth century a philosophical reinterpretation of the purposes of the First Amendment was on occasion used to limit its application to discussions of public policy only. The argument goes that the purpose of the First Amendment is not to protect individuals in the private gratification of self-expression but rather to protect the democratic policy-making process of society. According to this view, discussion of public affairs is what the First Amendment is all about. Views on politics are protected; other uses of speech, such as artistic expression, interpersonal relations, commerce, or just plain fun, deserve less or even no protection.

This philosophy, most coherently formulated by Alexander Meiklejohn, can still be detected in some recent opinions, though it is not the Court's general view.[57] Many *obiter dicta* note the special importance of protecting free public policy debate, while some opinions about three other areas—libel, obscenity, and commercial speech—deny that these deserve protection. However, in regard to commercial speech, the Court has drawn back from this tendency to pigeonhole speech into separate protected and unprotected categories.

The notion that commercial speech is unprotected came to the fore in Valentine v. Chrestensen in 1942.[58] In Jehovah's Witness cases the Court had earlier denied to communities the right to ban or license door-to-door canvassers.[59] Therefore Chrestensen, an advertiser, to get protection for his advertising, put a protest about police restrictions on one face of a handbill and an ad for a ship exhibit on the other. In a unanimous but offhand decision the Court declined to be fooled by this transparent device and upheld the ordinance restricting commercial handbills.

The decision, however, did not survive. It was at odds with precedents which held that free discussion does not lose its protection in a capitalist society by being conducted in profit-making media. Publishing corporations are "persons" entitled to the protections of the First and Fourteenth Amendments, the Court had ruled in 1935.[60] In several cases local authorities had tried to restrict, license, or specially tax the sellers of books or periodicals on the grounds that they were engaged in commercial enterprise. The Court consistently rejected the argument that publishers lose any First Amendment protection by being paid for their product.[61] After all, how will free discussion survive if it cannot earn its way?

This line of decisions in favor of commercial speech by publishers contrasts with the fact that there are innumerable restrictions on what businessmen may say in the course of an economic transaction, such as bans on false and misleading advertising, on offering stock without a registered prospectus, and on trying to influence employees' selection of a union. All of these restrictions have been treated by the Court as regulations on economic transactions, in which the speech is simply an integral part of an act like a purchase, sale, or contract. In the case of special taxes or licenses on commercial publishing, however, the aim of the activity is to produce information for public discussion. So for a while it looked as though the Court would try to segregate two domains: that of commerce for commerce's sake, in which speech would get no protection, and that of production of materials for public discussion, wherein the mere fact that profit is being made would not deprive the producer of protection.

A difficulty with this distinction, which led the Court to retreat from it, is that it forces the judiciary into case-by-case censorious review over advertisements to decide in each instance whether the content is a contribution to policy discussion or just a sales promotion. This is exactly like the situation the Court has gotten itself into with regard to obscenity. To take a hypothetical example, is a grocery ad a contribution to public health information or merely a sales pitch? What about a vitamin ad? And what about a birth control counseling ad? Where should the line be drawn?

The decision in Breard v. Alexandria in 1961 posed that dilemma in extreme form.[62] An Alexandria municipal ordinance prohibited all peddlers or canvassers from going to a door unless requested by the resident. Despite the long line of decisions upholding the right to canvass for religious or political purposes, this ordinance was upheld by the Court as applying to magazine salesmen, for the commercial element of their activity was held to be dominant.

But as time passed, the Court had second thoughts on the Chrestensen and Breard decisions. In 1959, Douglas, who had already dissented from Breard, recanted his vote in Valentine.[63] By 1975 in Bigelow v. Virginia the notion that commercial speech is unprotected largely passed from the scene. In that case a newspaper in Virginia, where encouraging abortion was illegal, had carried an ad for a legal New York abortion referral service. Faced with advertising of such social import, the Court concluded that "the Virginia courts erred in their assumptions that advertising, as such, was entitled to no First Amendment protection" and observed that the "relationship of speech to the marketplace of products or of services does not make it valueless in the marketplace of ideas."[64]

The doctrine that in the realm of commerce speech is unprotected was finally overruled in 1976 in Virginia State Board of Pharmacy v. Virginia Citizens' Consumer Council.[65] The politics of this case are as fascinating as its legal result. The pharmaceutical industry advertised its nonprescription pills and potions to the public at large but preferred to treat its prescription sales as a professional activity in which advertising was addressed to physicians alone, not patients, and thus price was not featured. Druggists had also persuaded many legislatures to prohibit drugstores from advertising the prices they charged for prescription drugs. Consumer advocates were unhappy with the result. Ralph Nader and his allies in the Virginia Citizens' Consumer Council demonstrated that consumers paid vastly different prices for the same product at different drugstores and often paid much more for the same product under a prescription than under a common name. They asked the courts to rule that the Virginia prohibition on drug ads containing prices violated free speech. At the same time, a different consumer organization, Action for Children's Television (ACT), was pleading with the FCC to stop drug advertising on programs that children viewed. Vitamin and medicine ads on television, it argued, encouraged children to believe that pill popping would solve their problems and would even start them on a path to hard drugs. They wanted advertisements regulated.

Various business groups also had conflicting interests in this situation, and each took advantage of the appropriate consumer group to make its case. The advertising industry was anxious to bury the legal category of commercial speech, for it deprived their activity of the protection of the First Amendment. They saw the suit by the Virginia Citizens' Consumer Council as an opportunity to do so once and for all. There could be no better way for the advertising industry to make its case for the validity of freedom to advertise than in a suit by the consumer movement.

The pharmacists, who wanted to continue their anticompetitive market practices by means of legislation against advertising, counterattacked by finding another part of the consumer movement to make their case for them. Attorney General Francis X. Bellotti of Massachusetts was prosecuting a drugstore that had advertised its ethical drug prices. Proceedings on that case were in abeyance pending the Supreme Court's ruling on the Virginia case, but in the meantime Bellotti did not sit still. He recognized that if the consumer movement could be mobilized in favor of regulations against televised drug advertising to children, it would make a case for exemption of commercial speech from the privileges of the First Amendment. Such regulations to protect children from the drug habit might look quite different to courts than regulations to protect pharmacists from informing consumers of the prices they were charging.

Bellotti, together with Action for Children's Television, petitioned the FCC to ban all drug ads from the air until 9:00 P.M. To cosign the petition with him, he got seventeen attorney generals from states that barred price advertising of ethical drugs. He had available the precedent of the "family viewing hour" (which no court had yet ruled unconstitutional), whereby the networks agreed to put on violent shows only after 9:00 P.M. when children are presumably in bed. The FCC called for testimony from psychologists and interested parties on the effects of drug advertising on children. On the First Amendment issue Bellotti argued that advertising enjoys little such privilege.

Bellotti's ingenious counterattack, however, was too late. Only a few weeks after the FCC sessions and presumably unaware of them, the Court issued its decision on the Virginia case, restoring to advertisers the right of free speech. Thus, since 1976, the Meiklejohn argument limiting the First Amendment to speech about public affairs has been rejected. Commercial speech is also protected. Expressions about sex that are designed to titillate are the main area in which the Court is still trying to carve out a zone of unprotected speech. In this area the Court is engaged in what will presumably prove to be the hopeless task of trying to stem the sexual revolution.

Rejection of the Meiklejohn theory that the First Amendment applies primarily to discussions of public affairs is also implicit in the Burger Court's refusal of special privileges for the institution of the press. Chief Justice Warren Burger has argued forcefully that the First Amendment is a protection for all persons, not just for a "definable category of persons," such as newspaper reporters, or for a special set of institutions, such as those engaged in the publishing business.[66] The state of Massachusetts had passed a law restricting corporations in using their resources in referenda campaigns. The Court in 1978 overturned this law as an infringement on free speech. Burger in a concurring dictum argued that if business corporations could be so restricted, then so could "media conglomerates" which are corporations too.

The *New York Times* was editorially appalled that "the First Amendment rights of, say, the *New York Times* may one day be judged legally no greater than those of General Motors." It approved Justice Stewart's remark in a 1974 speech that publishing is "the only organized private business given explicit constitutional protection."[67] But Burger and the Court majority do not read the Constitution this way. As Burger pointed out, when the government can decide who the press is and is not, this "is reminiscent of the abhorred licensing system of Tudor and Stuart England—a system the First Amendment was intended to ban from this country." The First Amendment protects a hippie dissident, a Jehovah's Witness passing out smudged tracts on the street, and the First National Bank of Boston in the same way and to exactly the same degree as it protects the *New York Times* or the *Atlantic Monthly.*

The commercial press is divided in its feelings about such rulings. Some reporters and editors claim that the institution of the press, because of its role in public affairs, has a special status that is constitutionally recognized. A police search of a newspaper office without a warrant, like that of the *Stanford Daily,* seems to them far more heinous than the search of a citizen's home without a warrant.[68] These advocates of press privilege believe that an employed reporter should be exempt from testifying before a grand jury as to what he or she has learned on the job, quite unlike a private citizen, political activist, social researcher, or volunteer who writes a handbill. Other reporters disagree; they oppose granting journalists special protections, such as so-called shield laws, which in some circumstances exempt reporters from testifying in court. The granting of privileges implies that the government has authority to exercise discretion in the matter.

Internationally, a similar debate centers on a UNESCO proposal for protecting the rights of journalists. American press spokesmen oppose such a convention because it would require government accreditation as a journalist in order to enjoy those rights. Government restrictions on reporting and writing by others than accredited journalists would implicitly be sanctioned. The danger in claiming special rights other than those that inhere in citizenship is that privileges which legislators give, they can easily take away.

The Court does not accept the view that the First Amendment applies in a special way to the press. Yet the Court, aware of the peculiarities of the social institution called the press, has stated the principles of the First Amendment in ways that do reflect a special regard for journalism. On the two issues of the refusal of reporters to identify sources and the inviolability of newspaper offices to search, the Court, while coming down on the side that denies newspeople any special privilege, has in its *obiter dicta* talked eloquently about the importance of the press.

The notion of differential protection for different kinds of speech is thus weakened but not dead. In an uncertain and wavering way the Court still grades different kinds of speech as more deserving or less deserving of its vigilance.[69] The Meiklejohn heritage can still be seen. Public affairs debates still have a particular sanctity in the eyes of the Court. But the notion that there is a domain of speech about public affairs that is free and a separate domain of private communication that is unprotected has been rejected.

Speech That Merges into Action

Finally, the Court has sustained government regulations controlling speech when that speech is an integral part of acts of a kind that the government may properly regulate. On this limitation of the First Amendment the Court has been unanimous. Black and Douglas, the First Amendment absolutists, questioned no more than did Frankfurter that an illegal act, even if words are used in its performance, acquires no exemption from prosecution by virtue of being entwined in speech. Indeed, for the absolutists the line between speech and action is the only legitimate basis for circumscribing First Amendment protections.

The distinction between speech and action underlies such commercial regulations as those on claims made in the course of purchases and sales, or on labor negotiations, or on stock offerings, or on false and misleading advertising. The fact that a person is saying something does not legalize a fraudulent money-making activity. Similarly, the government may act to prevent espionage, or revolution, or obstruction of justice, even though communication is used to perpetrate the act. "It rarely has been suggested," wrote Justice Black in Giboney v. Empire Storage and Ice Company, "that the constitutional freedom for speech and press extends its immunity to speech or writing used as an integral part of conduct in violation of a valid criminal statute."[70] In this case the action involved picketing by a union to force an employer to sign an illegal contract. The Court held that though picketing is speech, when used in an attempt to break the law, it becomes itself illegal.

These four strategies used by the Court to determine when government regulations on speech are legitimate despite the First Amendment show how close the Court has come to an absolutist libertarian position without at the same time embracing it. The history of twentieth century cases has been one of strengthening the First Amendment. The Court has given to freedom of speech a broad and growing protection. It has given it a preferred position in the nation's system of law. The Court has outlawed almost all censorship of publications. It has overruled repeated efforts to require permits for canvassing or for distribution of handbills, or to impose taxes on the sale or distribution of literature. It has protected picketing, soapbox speaking, and parading in public streets and parks. It has barred state governments from demanding membership lists of organizations or from requiring teachers to list their memberships. Although the Court has not quite accepted the absolutist argument that "no law" means "no law," it has gone along with Black's view that the command of the First Amendment must be read with the broadest scope. In various resounding declarations—such as, "Only the gravest abuses, endangering paramount interests, give occasion for permissible limitation"—the Court has come close to a libertarian view of the rights of citizens under the Constitution.[71]

Perhaps some day the dissents of Black and Douglas will acquire the same status in law as the dissents of Holmes and Brandeis have today. Given the composition of the Court in the early 1980s, it will not happen soon; but it may eventually. Whether it does or not, the Court has given a wide swath of protection to speech that is conducted in the traditional media of communication: print, meetings, parades, associations, and canvassing.

All these media have been invested with a well-defined set of immunities. Substantively, perhaps the most important of First Amendment protections for publishers is that against prior restraint, for this bars censorship. In reaction to ancient British abuses, the press received immunity not only from censorship but also from licensure, special taxes, and prosecution for criminal libel. In addition, by decisions of the Court as to what the First Amendment means, publishing now has defense against vague or harassing laws or enforcement procedures, and it receives special attention from the courts when it claims violations of the First Amendment.

But while the Supreme Court has avowed that the First Amendment applies to all expression and not just to political institutions and the press, what Congress and the courts have actually done is quite different with regard to freedom of speech in new media that were not used for discussion of public affairs in the seventeenth and eighteenth centuries. The record of vigilance for them is far more ambiguous.

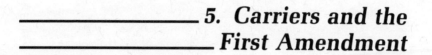

5. Carriers and the First Amendment

The mails, telegraphy, and telephony operate in America under very different legal principles from those that govern publishing. Carriers, because often they are monopolies, at least locally, are required to serve their customers in a quite impartial way. To this extent, though the law of common carriage differs from the law of press and platform, it is also designed to protect free expression. There is a certain amount of inadvertence in how this came about.

The Law of the Mails

A postal service needs to be universal. It should serve every crossroad and hamlet. But the volume of business at the end of the line rarely covers costs at those minor locations. Small branch offices, home delivery, and rural service are the bane of postal economics. In colonial America and through the first half of the nineteenth century the places where there was enough business to make a full-time postal operation profitable were the exception. Almost everywhere postal activity had to be an add-on to some other related business, as it is in rural general stores and local substations even today.

Newspaper publishing was one of the businesses that could be joined to postal service; they were in natural synergy. The post office could distribute the paper, and it was also a center to which news came, a fact that was important before there were any reporters and wire services. Just as important, before the First Amendment, was the fact that the postmaster had gotten his job because he was a political favorite, so the government was pleased to have him publish a paper. John Campbell, the postmaster in Boston, published the *Boston News Letter* from 1700. When he fell from favor and was replaced by William Brooker in 1719, the new postmaster started the *Boston Gazette*, and Boston for the first time enjoyed two competing papers.[1]

The most famous publisher-postmaster was Benjamin Franklin. In 1753 he and William Hunter, both already newspaper publishers, were named joint deputy postmasters general for the colonies, and when the Revolution came, Franklin became the new nation's first Postmaster General.

Postmaster-publishers had a clear conflict of interest, and sometimes they used their postal authority to discriminate against competing papers. Postmasters did not charge themselves for carrying their own papers. In this way a tradition got started of free or subsidized delivery for newspapers, which persists to this day in the form of second-class rates.

In the synergy between postal delivery and other businesses, like newspaper publishing, the cross-subsidization can go either way. Historically, postal service has often produced revenue for governments. The earliest mail systems were established not for the use of commerce or private citizens but for the purposes of empire. In the sixth century B.C. Cyrus, Emperor of Persia, had a system of couriers which resembled the "riding posts" of the Middle Ages, though it was more highly developed. The postmaster, an appointee of the Emperor, maintained a horse stable and had couriers stationed at the post. He received letters from the couriers of other posts and forwarded them by fresh couriers and horses to the next post on the way to their destination.[2] A similar Roman system extending from Scotland to Egypt was likewise for the use of the government only. After the fall of Rome, the next extensive European empire to create a postal system was Charlemagne's. The limits of a kingdom generally reached no further than the distance the king's messengers could ride to deliver commands and carry back reports.

Other institutions besides governments established their own systems of couriers. From the twelfth century, European universities, particularly the University of Paris, had such systems. From the fourteenth century the Hanseatic League had a similar arrangement for its own commercial use, not for the public's.

With the growing demand for communication, monarchs realized that money could be made on the side from postal service. In 1544 in Germany and Spain the monarch granted permission for private services, and shortly thereafter Charles V conferred a monopoly on the Count of Thurn and Taxis, which proved profitable.[3] In England, too, the posts were established as a national institution in the middle of the sixteenth century. The postmaster had a monopoly right to provide horses for travelers and also a duty to carry the king's messages to the next post. The king conferred the monopoly to ensure the financing of this network of messengers to meet his needs. The postmaster was not obligated to carry letters for private citizens; he did so to earn extra money.[4]

Service to the public was thus a byproduct, seen by both government and postmasters primarily as a source of added revenue. Little attempt was made to adapt the service to the needs of commerce or to make it universal. The routes went where government traffic required. Nonetheless for many postmasters and for governments the auxiliary commercial traffic, protected by a patent of monopoly, proved lucrative, and awards of such postal patents were a plum for royal favorites.

This tradition had its legacy in the American colonies, where until the 1820s the post office was run with an eye to making money for the government. Afterward national development goals were given primacy. With the emergence of this activist welfare orientation, which was riskier and less profitable, the system expanded, but its fiscal performance deteriorated. In 40 of the 61 years from 1789 to 1850, the post office was in the black, whereas in all the 130 years from 1850 to 1980, it recorded a profit only eight times. For a citizen in the eighteenth and early nineteenth centuries, getting the news was not the easy matter it is today. The arrival in port of a ship from abroad or at an inland location of the fortnightly post riders was an event of importance. From it flowed handwritten letters, talk in coffee houses, and exchange of newspapers that caught people up on the world's events.

The development of facilities for the dissemination of news was a recognized public need. Newspapers advocated road building and post office extension. The inclusion in the Constitution of a power for the new federal government to establish post offices and post roads was neither a routine nor a trivial matter. Public works such as canals and roads were key developmental issues of the day. Until the telegraph, the post was the sole vehicle for transmitting news, and its importance to public life in the newly developing nation was everywhere recognized. The feelings of that day about the post and the press were expressed in 1817 by Senator John C. Calhoun: "Let us conquer space. It is thus that . . . a citizen of the West will read the news of Boston still moist from the press. The mail and the press are the nerves of the body politic."[5]

This was the view that led the Constitutional Convention to give Congress the power to set up a postal system. They recognized its usefulness to commerce but were equally aware of its value to the press. The twentieth century notion that the proper relation between government and the press is one of arm's-length adversaries has no roots in the thinking of the founding fathers. Their belief in the importance of the press not only led them to insist that Congress pass no law "abridging the freedom of . . . the press" but also persuaded them to subsidize that press.

When Thomas Jefferson in 1800 was elected President, he used his power to promote a newspaper. He urged the Philadelphia publisher Samuel Harrison Smith to move to Washington to issue an organ, *The National Intelligencer*, favoring the new administration. It was given financial support by being named printer to Congress. As the tides of politics shifted during the first half of the nineteenth century, various other papers were favored with government ads and printing contracts in return for serving as a presidential mouthpiece.[6]

To help build up a press seemed to politicians of the day to be a proper and useful task for government. They saw no contradiction between Article 1, Section 8, of the Constitution, which allowed the government to set up a postal system, and the First Amendment. On the contrary, the consensus that the postal service should aid the press was expressed in the provisions of the first Post Office Act of 1792, which set the postage for newspapers at one cent each and allowed every newspaper editor to send one copy of the paper free to every other newspaper in the country, thus providing at public expense the primitive equivalent of a news service.

Granting franking privileges or concessionary rates to publications was one of the developmental policies that cut into revenue. This privilege, said a congressman in 1874, is "one of the many ways under our form of government in which we educate our people." In much of the world, free or subsidized newspaper dissemination continues in various forms today.[7]

In the United States during most of the nineteenth century, the newspaper receiver, the sender, and the government shared the mailing costs.[8] The newspapers paid the intercity postage. If the subscribers wanted the paper delivered to them from the local post office, they generally paid the postman for that extra service; otherwise they picked up the paper at the post office. Window service was often free. A law of 1845, repealed after two years, provided for free local newspaper delivery at the post offices within thirty miles. In 1851 free delivery was again voted, this time within the county of publication, and rates within one hundred miles were set lower than beyond. The purpose of these laws was to protect local papers against metropolitan competition.[9] The free exchange of papers among newspaper publishers continued. When in 1875 this franking privilege ended, very low second-class mailing rates continued.

Similar postal policies were followed for books and magazines. In 1851 books were first admitted to the mails, and in the next year they were given a concessionary rate. Magazines in 1852 were given the same rates as newspapers. The result was, on the one hand, a vast growth of the press and, on the other, of postal deficits. Thus congressional policy in the nineteenth century supported the press and, in particular, supported small-town papers rather than metropolitan ones.

By some measures the delivery of newspapers was the main business of the post office. There is no exact way to equate the significance of a half-ounce letter with so many pounds of newsprint, but the importance of the system to newspaper publishing is shown by the fact that as early as 1794 newspapers made up seven-tenths of all mail matter. In 1832 printed matter was said to be fifteen-sixteenths of the total weight of the mails, though it provided only one-ninth of the revenue.[10] In 1901 second-class mail was three-fifths of the weight but paid only 4 percent of the postal revenue. By 1909 the overall post office deficit was $17,500,000, but the second-class deficit alone was $63,000,000; the rest of the system made money. From the 1790s until 1918, with the exception of a temporary increase during the War of 1812, Congress did not permit any increase in postage rates for newspapers.

Rural free delivery was another step toward social goals for the postal system at the expense of the quest for profits. Postal service became a right of every citizen, no matter how remote the residence and how little revenue might be brought in. The postal service was an attribute of nationhood, not a business. It was promoted as a prerequisite for free debate and public discussion.[11]

Press Freedom

An intimate relationship exists between the carrot and the stick. A government that gives subsidies can intimidate those who become dependent on them by threatening their withdrawal.[12] From that fact those who fear government coercion can draw the conclusion that government should serve only night watchman functions and not be relied on for favors. That was a Victorian laissez-faire conclusion, not one prevalent in the eighteenth century. Alexander Hamilton was a mercantilist, not an advocate of laissez faire. James Madison was not William Graham Sumner. So in embryonic form an issue was endemic in the constitutional provision for a postal system. The service it could offer contained the potential for favoritism being used for coercion. The First Amendment denied Congress the power to abridge the freedom of the press, but the government could conceivably manipulate the press through the postal services it offered.

In 1792, pursuant to the postal power, Congress granted a monopoly to the post office for the carriage of mail. In 1825 Congress reaffirmed the monopoly, forbidding the commercial transport of letters, packages, or other mailable matter "between places, from one to the other of which mail is regularly conveyed."[13] By the eighteenth century, the idea of a postal monopoly was already well established in British practice, and was familiar and natural to the Americans.

In Europe the monopoly principle is usually called the "third-party rule." According to this rule, a writer, party A, may hand a message to party C, to whom it is addressed, but a private third party, B, may not convey the message from A to C. The third-party rule is applied in Europe to telephony, telegraphy, and computer networks as well as to the mails.

This monopoly rule was contested in Puritan England. In 1659, just fifteen years after Milton's *Areopagitica* attacking the state's claim of the right to license printing, one John Hill wrote a pamphlet entitled "A Penny Post: or a Vindication of the Liberty and Birthright of Every Englishman in Carrying Merchant's and other Man's letters, against any Restraint of Farmers, etc." Then in 1682 William Dockwra established a private delivery service in London, with postage costing a penny, five hundred collection boxes, hourly collections, and six to ten deliveries daily.[14] The government prosecuted Dockwra, secured his conviction, and assimilated his service, without its cheapness and efficiency, into the post office. In 1708 Charles Povey tried to start a halfpenny post in London, but it too was promptly suppressed. Thereafter the postal monopoly in Europe remained essentially unchallenged until the development of electronic mail.

In America the phrase "third-party rule" has not been much used, but the same issue, called "postal monopoly," has caused controversy. No stronger instrument of censorship can be imagined than a monopoly on the means of delivery. As a result, there is substantial tension between the postal power and the First Amendment.

Two constitutional issues can be distinguished: the legality of a communications monopoly and the right of a carrier to discriminate in what it will or will not carry. The intersection of these two questions defines four policy positions. The most restrictive affirms the right of the government to set up the postal service as a monopoly and also affirms the right of such a monopoly not to carry certain classes of material, such as obscene matter, lotteries, or false and misleading advertising. This position in effect denies to purveyors of the excluded material any means of delivery. The most libertarian position denies to the government the right to give any agency the exclusive right of message carriage and also denies to any delivery agency that the government establishes any power of discrimination in the content of what it carries. This position limits the government's role in delivering communication to a purely ministerial one with no censorship power. Of the two intermediate positions, one denies to any delivery system a monopoly but allows any of the plural delivery systems to restrict their service to certain classes of message. The other intermediate position permits the government to establish carrier monopolies in some situations but obliges such monopolies to carry any message that a customer gives them. These policies can be charted:

	Monopoly allowed	*Monopoly forbidden*
Discrimination by content allowed	most restrictive	intermediate
Discrimination by content forbidden	intermediate	most libertarian

The policy represented by the most restrictive combination was early recognized as incompatible with the First Amendment. Both the Supreme Court and the Congress rejected it. But neither did they take the most libertarian position. Historically, they have oscillated between the two intermediate positions. At times they have authorized certain monopolies in communication carriage but as a corollary have denied that the carrier had discretion about the contents of what it carried. On other occasions the Court and Congress have allowed the exclusion of certain kinds of materials from delivery by a public carrier, but as a corollary to this policy have denied the carrier the right to be the exclusive means of delivery. Inadvertently, some very restrictive policies have resulted from the combination of the two intermediate positions.

In 1836 President Andrew Jackson, concerned at the upsetting effect that the distribution of antislavery propaganda might have in the South, urged Congress to ban such matter from the mails. An ad hoc committee of the Senate chaired by Calhoun, America's classic defender of slavery, rejected the President's request because, in his view, Congress had no such power: "If it be admitted that Congress has the right to discriminate . . . what papers shall or what shall not be transmitted by the mail, [it] would subject the freedom of the press, on all subjects, political, moral, and religious, completely to its will and pleasure." Calhoun, in short, chose the intermediate position that forbids discretion about content but allows monopoly. For Calhoun, a spokesman of the slaveholding South, this conclusion was acceptable because then, before the Fourteenth Amendment, it was argued that the states could ban such matters as antislavery propaganda. However, the federal government, under the First Amendment, could not use its postal powers to do so.[15]

In line with Calhoun's view on the legitimacy of the postal monopoly was an 1851 Supreme Court case, United States v. Bromley.[16] Bromley, who had been carrying messages for a fee, in his defense unsuccessfully challenged the constitutionality of the monopoly provision of the postal law. The Court in its decision implicitly recognized the First Amendment as a more important clause of the Constitution than the postal power; if these powers were in conflict, the First Amendment had precedence. The Court did not contest the defendant's claim that, in carrying messages, he was exercising his basic freedom of speech, of which he could not be deprived just because the Constitution also allowed Congress to establish post offices and post roads. The Court based its decision on a different ground. The issue was not between the First Amendment and the postal power but between the First Amendment and the revenue power. The purpose of the monopoly provision was to protect the revenues of the federal government from the competition of private carriers who would deprive the government of the income from postage stamps.

Among the various clauses of the Constitution, the Court has given different weight to different ones. Historically, some clauses have been interpreted narrowly, and others have been made to carry increasing freight. The Court, in giving to the First Amendment a preferred position in the law of the land, has found it to represent constitutional concerns "of far greater importance than the transportation of the mail," but it is not the only constitutional provision given great weight and broad construction by the Court. The revenue power is another. Indeed, the so-called elastic clause refers specifically to the use of the revenue power: "The Congress shall have the Power . . . to lay and collect Taxes, Duties, Imposts and Excises, to pay the Debts and provide for the common Defense and general Welfare of the United States."[17] Congress was not given the power to legislate for the general welfare as such, but it was given the right to tax and spend for that broad purpose. And so the government's case for its postal monopoly, when challenged in court, was that "the law under which the suit was brought is a revenue law, and is to be liberally construed."[18]

The key issue in Bromley was therefore whether the postal monopoly was based on the postal or revenue powers of Congress. If the former, a monopoly could not be defended in the face of the First Amendment; if the latter, it could be. The decision marshaled evidence that the issue was the government's revenue. The income from postage stamps was pooled in the Treasury Department as part of the general income of the government, as well as being potentially essential to the viability of the postal service. The postage fees were even called "a tax on mailable matter" in the postal act. That the act was therefore "a revenue law," the decision concluded, "there would seem to be no doubt."

The term "cream skimming" had not yet been invented, but the need to protect the government from cream skimming was implicit in this decision. Just as AT&T more recently protested that specialized carriers would undercut the health of the system by competing for the most profitable services while leaving the principal carrier with the obligation of providing the unprofitable ones, so too the American government in the nineteenth century was concerned that competition would impair its postal revenues and leave it with only the unprofitable part of the business.

Another challenge to the postal monopoly occurred in 1877 in Ex parte Jackson. One A. Orlando Jackson was jailed for conducting a lottery through the mails. He pled that the post office, by the nature of its monopoly, had no right to bar anything from the mails. Its role was simply to carry whatever was given it, not to decide what was or was not mailable.[19] This time the Court swung to the other intermediate policy position shown in the chart. The Court rejected Jackson's argument that Congress could not decide which kinds of matter were mailable and which were not. Justice Stephen J. Field found it reasonable for Congress to decide that lotteries were immoral and thus not to be sustained by postal service. However, he recognized the danger of censorship thus created. He resolved the dilemma by arguing that whatever matter the post office would not carry, citizens were free to deliver in any other way they chose. Field opted for the monopoly-forbidden, discretion-allowed position. The postal monopoly could not be used to bar transmission. Any service that the monopoly chose not to provide, others could provide at will: "The right to designate what shall be carried necessarily involves the right to determine what shall be excluded. The difficulty attending the subject arises . . . from the necessity of enforcing them consistently with the rights reserved to the People, of far greater importance than the transportation of the mail."

Among these rights "of far greater importance than the transportation of the mail" were both freedom of speech and protection against unreasonable search and seizure. The latter was the crucial point for sealed first-class material: without a warrant, the post office had no right to open letters and so had no way of excluding unmailable matter. With regard to open printed matter, the contents of which could be easily established, the crucial point was the First Amendment. That superior norm could not be maintained if some items were undistributable: "Nor can any regulations be enforced against the transportation of printed matter in the mail, which is open to examination, so as to interfere in any manner with the freedom of the press. Liberty of circulating is as essential to that form as liberty of publishing; indeed, without the circulation, the publication would be of little value. If, therefore, printed matter be excluded from the mails, its transportation in any other way cannot be forbidden by Congress."[20] Justice Field rejected Calhoun's argument that Congress under the First Amendment had no authority to determine what type of content was mailable.[21] Field believed Congress had such power because it did not possess "the power to prevent the transportation in other ways, as merchandise, of matter which it excludes from the mails." The Court thus restricted the scope of the postal monopoly rather than of regulations on what the mails would carry.

Ex parte Jackson was not the only judgment restricting the scope of the Bromley decision and reflecting the American antipathy to monopoly. In 1910, for example, a federal appeals court decided that the term "mail" in the law referred to correspondence and that the government had no monopoly on parcel delivery.[22] Nonetheless, in its limited domain the postal monopoly came to be accepted as the law of the land, and so did the right of Congress to determine what was mailable.

Cautious as Congress and the courts had tried to be in handling cases on the limits of postal power, the lack of a clear position on the dilemma between monopoly and nonmailability led by the 1930s to the worst of both worlds. Decisions swung between the two intermediate positions: sometimes monopoly was justified by the absence of post office control of contents, and sometimes exclusion of nonmailable matter was justified by limiting the post office monopoly. In this confusion the central dilemma was lost sight of, and precedents were established both for monopoly and for control of what could be mailed. The outcome for a period was a legal practice in the most restrictive corner of the fourfold table. Beginning in the Civil War and lasting unchecked through World War II, an insidious and familiar process was at work: the gradual expansion of regulatory power.[23] This happened in a reform era when moralism and meliorism were in the air. Proponents of all sorts of good causes, like temperance, labor unions, and women's suffrage, were turning to government to set wrongs right.

The use of the postal system for reform began with innocent-seeming provisions. In 1865 Congress passed an act barring obscene publications from the mail. In 1868 and 1872 Congress barred illegal or fraudulent lotteries from the mails, and also scurrilous epithets and devices on postcards. Until those laws, mailable matter had been defined only by such aspects as size, weight, shape, and wrapping.[24] It seemed perfectly legitimate for the federal government to help states enforce their laws barring lotteries by denying to lottery promoters the privilege of federal assistance, particularly since under the supremacy of federal powers, states could not interfere with the carriage of the mails. The federal government had previously been acting as an accomplice in the violation of state laws.[25]

The law on lotteries was expanded four years later to ban all lotteries from the mails regardless of their legal status in a state. The moral reformism of the age was encouraging new governmental interventions at the same moment that it was discouraging old practices of patronage used for the manipulation of the media. The net result of these opposite trends was hardly favorable. In 1888 the ban on obscene publications was extended to the contents of sealed letters. In 1890 newspapers with lottery ads were banned from the mail. At the turn of the century letters and circulars used in other kinds of fraudulent schemes, such as chain letters, were also held to be unmailable.[26]

In 1908, at President Theodore Roosevelt's urging, the Attorney General interpreted seditious content of an anarchist journal as indecent, and in 1911 Congress amended the law to define indecency to include "matters of character tending to incite arson, murder, or assassination." In 1912 prize fight films were barred from the mails. In 1933 it was made illegal to use the mails for selling securities unregistered with the Securities and Exchange Commission. The gradual accumulation of more and more restrictions on the mails, once the principle was accepted that some things could be declared nonmailable, shows the wisdom of Justice Joseph P. Bradley's remark in 1886: "It may be that it is the obnoxious thing in its mildest and least repulsive form; but illegitimate and unconstitutional practices get their first footing in that way, namely by silent approaches and slight deviations from legal modes of procedure."[27] By now the list of things that have at some time been declared unmailable includes:

Objects of excessive size
Inflammables or explosives
Infernal machines
Poisons
Perishable goods
Disease germs
Intoxicating liquor
Liquor ads when liquor was illegal
Solicitations that may appear to the receiver to be a bill
Unsolicited contraceptives
Unsolicited ads for contraceptives
Motor vehicle master keys
Prize fight films
Fraudulent matter
Lotteries
Obscene matter
Sexually oriented ads
Pandering ads

Unordered merchandise
Security offerings other than by an approved prospectus
Matters inciting to arson, murder, or assassination
Material obstructing conscription

Many of these prohibitions seem legitimate. One could hardly deny to the post office the right to refuse oversize packages, or bombs, or unrefrigerated fish. Nor is any First Amendment issue raised by the ban on unordered merchandise, motor vehicle master keys, or unsolicited contraceptives, though some Fourth and Fourteenth Amendment procedural issues may be raised by whether postmasters may look inside and themselves decide the legality of the object that is offered for mailing. But in the ban on fraudulent representations the issue is joined. The question is whether the post office has any business to decide what statements constitute fraud, or whether this is a matter for the courts. And in the case of obscene matter the fundamental question is again whether, under the First Amendment, any agency of the federal government may properly restrict it.

The defense of all of these regulations on mailability was, as in Ex parte Jackson, that the service denied to the sender was only one among various alternative means for sending a document. In sustaining Coyne, a postmaster who had refused mail from an organization apparently engaged in a fraud, the Court in 1904 said that the postal service "really operates as a popular and efficient method of taxation. Indeed, this seems to have been originally the purpose of Congress. The legislative body, in thus establishing a postal service, may annex such conditions as it chooses."[28] As evidence for this decision that the posts are not an indispensable means of delivery or a necessary adjunct to civil government, the Court noted that earlier rates had been so high and delivery so uncertain that the mails were not very important. Only when stamps were introduced in 1845 were the letter rates reduced to five and ten cents. So the mails were a convenience, not a necessity, and Congress could set conditions on their use.

This line of argument occurred repeatedly. In 1912 Congress required an elaborate statement of ownership as a condition for getting second-class rates. The Lewis Publishing Company claimed that forcing it to reveal its owners violated its First Amendment rights—a contention the courts have upheld in some nonpostal contexts.[29] In this case, however, the Supreme Court held that since first-class mail was still available, a registration requirement for second-class rates was unexceptionable. The Court rejected a government claim of broad and arbitrary power over the classification of mail but sustained the questionnaire as asking for information pertinent to the privilege of sending subsidized mail. The crucial point, once again, was that there were alternative means, albeit more expensive ones, for distribution by those who did not choose to register.

In a pair of dissents, which are now the law, Justice Holmes took exception to this line of reasoning. He challenged the premise that alternatives were available, maintaining that the mails are a unique means of communication, and thus the government under the First Amendment may not arbitrarily choose to make them available or withdraw them at will. The first case in 1921 concerned Postmaster General Burleson, who denied the socialist paper *The Call* second-class mailing rights on the grounds that it was seditious and so not entitled to special privileges.[30] The Court upheld him, but Holmes dissented, arguing that while the government was not obligated to provide mail service of any kind, if it did provide it, it must do so without discrimination: "The United States may give up the Post Office when it sees fit, but while it carries it on the use of the mails is almost as much a part of free speech as the right to use our tongues."

A year later, Holmes, joined by Louis Brandeis, again dissented in the case of a certain Leach who was accused of engaging in patent medicine fraud. The Court denied him the use of the mails for that activity. But Holmes said, "When habit and law combine to exclude every other [means] . . . the First Amendment in terms forbids such control of the post as was exercised here."[31]

This reasoning resurfaced in 1941 in a decision of the District of Columbia Court of Appeals. In Pike v. Walker, Postmaster General Frank C. Walker conceded that he had given no notice and conducted no hearing before issuing a fraud order banning Pike's mail from delivery, but he claimed that he did not need to because "no individual has a natural or constitutional right to have his communication delivered by the postal establishment of the government." A postmaster's action, he claimed, was discretionary. The Court rejected this: "Whatever may have been the voluntary nature of the postal system in the period of its establishment, it is now the main artery through which the business, social, and personal affairs of the people are conducted . . . Not only this, but the postal system is a monopoly which the government enforces through penal statutes forbidding the carrying of letters by other means."[32]

Finally, four years later in Hannegan v. *Esquire* the Supreme Court embraced the Holmes-Brandeis viewpoint, overturning in the process the tradition from Ex parte Jackson through Coyne to Milwaukee Publishing Company. Postmaster Hannegan had tried to deny second-class mailing privileges to *Esquire* on the grounds that this then somewhat raunchy magazine was not serving the public interest, the purpose for which Congress had granted low rates to periodicals.[33] Justice Douglas, speaking for the Court, said that this was not a matter for the Postmaster General to determine.

Thus since the mid-1940s the law has been that the postal service is a de facto and de jure monopoly, and as such, it must, without discrimination, be open to all who would express themselves in ways that the Constitution protects. This leaves only a procedural question as to how the post office should go about separating out those things, like obscenity, to which such protection is denied. The main procedural issue is the respective roles of postmasters and courts. If a magazine with sexy pictures is posted, is a postmaster qualified to decide whether it is obscene or not? And if the decision must be made by authorities other than the local postmaster and with careful procedures, what are the limits on permissible delay?

The limits on the postal service's discretion were spelled out in 1971 in Blount v. Rizzi. Congress had specified that there be no interference with mailed matter without an administrative determination of obscenity by the Postmaster General. The Supreme Court, however, ruled that this procedure was fatally flawed because a judicial review would occur only if the defendant appealed. The Postmaster General could act independently without obtaining prompt judicial judgment of the material's obscenity. Only the judicial branch, the Court concluded, has "the necessary sensitivity to freedom of expression." So where First Amendment considerations are at stake, "only a procedure requiring a judicial determination suffices to impose a valid final restraint," and the route to that judicial determination must be prompt and be sought at the initiative of those who would regulate.[34]

Through such cases since World War II the Court has stepped back from the situation that it created when it accepted both a postal monopoly and the right of the federal government to determine what is mailable. The single most important case was Lamont v. Postmaster General in 1965.[35] A law required the post office to hold up "communist political propaganda" mailed from abroad and, before mailing it on, to ask the addressee whether he or she wished to receive it.

The post office had sent Corliss Lamont an inquiry as to whether he wished to receive a copy of the *Peking Review* that was addressed to him. Rather than reply, Lamont sought an injunction against enforcement of the law on grounds of unconstitutionality.[36] The Court, in a rare unanimous decision on First Amendment law, sustained him.

This case was important in several respects. For one thing it declared the relevance of the First Amendment not only to the publisher of information but also to the recipient. "It is true," Justice William J. Brennan noted in a concurring opinion, "that the First Amendment contains no specific guarantee of access to publications. However, the protection of the Bill of Rights goes beyond the specific guarantees to protect from congressional abridgment those equally fundamental personal rights necessary to make the express guarantees fully meaningful . . . The Government argues that, since an addressee taking the trouble to return the card can receive the publication named in it, only inconvenience and not an abridgment is involved. But inhibition as well as prohibition against the exercise of precious First Amendment rights is a power denied to government." Since Lamont, the law seems quite unambiguous: the postal authorities may not refuse to deliver disapproved propaganda.[37]

Thus the Supreme Court, after a century of neglect, has once more brought the First Amendment to bear on the mails. It retains the nineteenth century rulings permitting a postal monopoly, although those rulings were based on obsolete fiscal arguments. But the Court now firmly insists that if the government wishes to exercise such a monopoly, it must open the mails to all viewpoints without discrimination. While the Court allows Congress to exclude from the mails unprotected material such as obscene publications, speech that is protected by the First Amendment may not be excluded if the mails operate at all; and "procedures violate the First Amendment unless they include built-in safeguards against curtailment of constitutionally protected expression, for Government 'is not free to adopt whatever procedures it pleases . . . without regard to the possible consequences for constitutionally protected speech.' "[38]

On the whole, the Supreme Court has now established a situation in which freedom of the mails is substantially protected. It reached this point by a tortuous route, in which censored mail almost came into existence. Until the *Esquire* case, the scope of federal power over what could be mailed grew without visible bounds. In the end, the

Court put its finger in the hole in the constitutional dike. The hole, however, was one that it had drilled itself when it accepted the practice of postal monopoly and then also affirmed the discretion of Congress over what the monopoly could be used for.

The Law of the Telegraph

In most countries today telegraphy is under the postal authorities. The third-party rule applies. With only limited exceptions, no private company may accept a message from outside the company and transmit it by telex or leased lines to another party.[39] For example, airlines are often prohibited from using their reservation network to book a car or hotel room for a passenger. In America, however, the history of telegraphic law is quite different and more complex. Telegraphy was never assimilated by the post office. It evolved in parallel.

American telegraph law might seem to have been modeled on postal regulations, but postal law was at most the grandfather of telegraph law rather than the father; postal law's impact was indirect. Telegraph law was modeled rather on the law that had grown up in the nineteenth century to regulate the new railroads. Railroad law got some of its concepts from the practices regarding post roads, so indirectly the postal influence is there, but the immediate model for telegraphy came from the evolving concept of a railroad common carrier.

As a consequence, the First Amendment is almost undetectable in cases concerning telegraphy. It might seem odd that when a new technology of communication came into existence, the courts did not perceive it as an extension of the printed word, sharing the same significance, the same infirmities, and the same need for protection as the press whose liberties the courts had so sedulously guarded. The reason for this dim perception of telegraphy was that the early telegraph carried so few words at such a high cost that people thought of it not as a medium of expression but rather as a business machine.[40] The computer suffered the same misperception a century later.

Evidence of this perception of telegraphy is the fact that in early telegraph cases, as later in early radio cases, the courts concluded that the federal government had authority to regulate under the commerce power.[41] The power of the federal government to regulate "commerce among the states" has never been seen as giving Congress the right to regulate the press, even though papers are sold as an item of commerce across state lines; the First Amendment supervenes. So the perception of the courts that telegrams were a proper object of federal regulation could only mean that they were in the eyes of the courts more analogous to packages than to newspapers.[42]

That such a misperception should have persisted beyond the Civil War is extraordinary, for in the 1850s and 1860s press use of the telegraph had boomed. In the very first years of the telegraph it was easy to misperceive it as a business device rather than a publishing device, for neither the press nor the telegraph companies were sure that they were made for each other. The synergy between them became obvious only after some years of jockeying for position and of testing of who was to be in control.

Even before the Washington-New York telegraph line was completed on June 7, 1846, newspapers had started experimenting with the telegraph. On July 4, 1845, a teleconference of editors was held to consider the use of the telegraph on the new New York-Albany-Buffalo line. On May 12, 1846, the *New York Tribune* started the first column of telegraph bulletins.[43] They were as brief and condensed as possible because the cost was so high. The outbreak on the next day of the Mexican War was the first hot news story sped by telegraph, and it triggered the interest of many other papers. So editors were interested from the start, but they were not yet ready to generate much traffic for reasons of cost. The New York-Boston line and the Pittsburgh-Louisville line both charged fifty cents for ten words, although some lines charged as little as two cents a word.[44]

The press was powerful, useful for public relations, and potentially an important customer of telegraph systems. It was able, using its bargaining power, to negotiate press tariffs at one-third to one-half of normal rates.[45] Even so, individual newspapers could not afford much telegraphic reporting. If the press were to use the new facility, it would have to do so cooperatively; newspapers would have to share the same wire report. This led to the concept of a news service. It also led to a battle as to who should provide the news service: the telegraph company or the press. Should the telegraph companies be information distributors, like the radio networks of today, or should they just be carriers, like the modern phone system, passively transmitting what the press organizations gave them? Telegraph companies tried to go into the business of offering news services, believing —as do cable television companies today— that their network gave them a natural basis for publishing. But the press fought off this effort, in the end coming to control their own news services.

The battle between telegraph companies and the press was fought in both the United States and Europe. In the United States in the 1840s, telegraph companies sold news services to papers for $5 to $10 a week.[46] They had bright young telegraphers in every community who doubled as reporters, and they had the national lines for distribution. When the Associated Press (AP) was formed in 1848, the head of the Portland, Maine, to New York telegraph companies, Francis O. J. Smith, chose to fight it for control of the international news business. He declined to carry the AP's traffic.[47] This loss would have been important under any circumstances, but it was particularly important before the transatlantic cable, because ships reached Halifax with the news from Europe before they reached New York. With access to the Portland-New York telegraph leg cut off, AP stood to lose the race to deliver European news from the first port of call. Newspapers ran vehement editorials about the dastardly tactics of Smith, their theme being the evil of monopoly. AP's response was aggressive and successful: it pieced out and filled in a line from New York to Portland parallel to Smith's. AP fought a good fight, but not in terms of any abstract principles of free press. Indeed, AP made contracts with various telegraph companies and provided some of the capital for them, in return for which AP received priority over its competitors in the transmission of news.[48]

The same sort of battle in Europe was again won by the newspapers. Three large press agencies were involved. Havas had been established in Paris before the electric telegraph and used pigeons to speed its news. Wolff's was formed in Berlin immediately upon the opening of the Berlin-Aachen telegraph line in 1849. Reuters was formed in London in 1851 as a result of the cable from Calais to Dover.[49]

Neither these organizations nor the telegraph itself was welcomed by the dominant papers. Mowbray Morris, manager of the London *Times*, wrote his Berlin correspondent in 1853, "I do not confide much in the telegraph and I would it had never been invented."[50] Running a paper with 60 percent of the London circulation and most of the foreign news, he did not welcome an equalizing device that served all papers. The provincial newspapers, on the contrary, found the telegraph a godsend. To a large extent they got their news from services provided by the telegraph companies.

Several British telegraph companies did as their colleague Smith had done on the New York-Portland line and refused to carry stories from correspondents employed by papers. Large papers like the *Times* resisted this pressure. Morris rejected a proposal to join a Berlin telegraphic news service on grounds that it was the paper's duty "to obtain authentic intelligence from every quarter of the globe . . . and for that purpose they retain correspondents whose duty is to supply them with such intelligence. These gentlemen are . . . responsible for the information they give and it is this responsibility which constitutes the chief security of their employers." The Berlin proposal would "substitute for the individual responsibility of a gentleman specially retained to serve a particular journal the absence of all responsibility necessarily implied by the very constitution of the institution in question . . . We would much rather remain in ignorance of information conveyed in such a manner."[51] The provincial newspapers, too, were not well satisfied when they found themselves dependent upon such companies as the Electric Telegraph Company, whose Intelligence Department, with a staff of reporters and editors, served 120 papers by 1854, and which by 1859 made an exclusive contract as a source for Reuters.[52]

The discontent of the papers at their dependence on the telegraph companies led them to campaign in 1869 for the nationalization of telegraphy in Britain. The Association of Proprietors of Daily Provincial Newspapers attacked the "despotic and arbitrary management" of the telegraph companies, and they announced the formation of their own cooperative press service. After an investigation, Parliament agreed to turn the domestic telegraphs over to the post office, with the injunction that the post office was to have no part in collecting news.[53]

Press use of telegraphy was growing. In 1854 the British telegraph had been just a business machine: 50 percent of its messages related to the stock exchange, 31 percent were commercial, 13 percent dealt with family affairs, and the press disappeared in "other."[54] But by 1875 the now nationalized telegraph service was transmitting 220 million words a year for the press at an average rate that had fallen to four pence per 100 words.[55] On this business, with a revenue of 50,000 pounds, the post office was losing 20,000 pounds. A major effect of such reduced rates was to bring into being the half-penny evening newspaper. Two such newspapers existed in 1870, seventy of them in 1893.[56] In 1907 the press telegraph deficit was 223,000 pounds, against a revenue of only 146,000 pounds; by 1912, when press messages were 11 percent of the British traffic, losses were much larger. But shortly thereafter the deficit problem began to disappear, along with reliable figures about the number of words of press traffic, for in the twentieth century the press services went over to private lines rented for a flat fee on which no message count was kept. By 1918 the press messages had fallen to 7 percent of the total, and eventually they almost vanished. In 1955 press revenue to the telegraph service was only 32,000 pounds for 373,000 messages; the rest of the press messages were on private lines.[57]

In the United States, where telegraphy was carried on for profit, press telegraph rates never fell as low as in Great Britain. Press traffic nonetheless grew. Morse's original telegraph line from Washington to Baltimore, which had been built under the auspices of the post office with a congressional appropriation, was sold to private interests in 1847. By 1851 there were fifty telegraph companies, most of them licensed under the Morse patent.[58] In 1856 a large number of companies were merged into the Western Union Company, which in 1866 absorbed two other large companies. In 1861 Western Union spanned the continent from coast to coast. By 1869 it carried 370 million news words for a total revenue of $884,000, or about one-fourth of a cent a word.[59] Western Union had an exclusive contract with AP and other news services at the expense of competing telegraph companies. In return, Western Union gave AP stories priority. By 1880 press telegrams were 11 percent of the total. But in 1879 AP leased its first private line, linking New York, Philadelphia, Baltimore, and Washington. In 1884 it leased its second line, from New York to Chicago, for which it paid $100,000 a year, compared to a total AP billing by Western Union in 1880 of $393,000.[60]

The special characteristics of a common carrier were soon imposed upon American telegraph companies. In 1866 Congress included in the Post Roads Act valuable privileges for telegraph companies so as to encourage the growth of the system. It authorized telegraph companies to run their lines freely along post roads and across public lands. It also permitted them to fell trees for poles on public lands gratis. To be eligible for these privileges, the companies had to provide service like a common carrier, namely to all comers without discrimination. By 1893 the Supreme Court ruled that telegraph companies, though not strictly common carriers, were similar: "Telegraph companies resemble railroad companies and other common carriers, in that they are instruments of commerce" and thus are required to provide service without discrimination.[61]

Besides that obligation, a number of other practices are characteristic of common carriers. Under the law that had evolved for railroads, they may not ask their customers to waive their rights, for these rights arise from common or statute law and not from the contract for service; the small print on the back of the ticket is not necessarily binding.[62] Common carriers are also often required to obtain a license to operate. The license, however, does not necessarily make them a monopoly; taxis are common carriers, legally obliged to pick up whoever hails them, but there are many competing taxis. And common carrier rates are often regulated.[63] Under the Communications Act of 1934, which defined both telegraph and telephone companies but not broadcasting stations as common carriers, telegraph and telephone rates are regulated by the FCC, but the rates that broadcasters charge their advertisers are not.

Imposition of common carriage requirements on the telegraph companies was a middle-of-the-road response to Western Union's monopoly abuses. The behavior of Francis Smith was not unique in telegraph history. When Henry George established a newspaper in San Francisco, the telegraph company there put it out of business by refusing it wire service.[64] More radical solutions than common carriage were advocated by some telegraph critics, namely the post office creating a competing postal telegraph service or even making telegraphy a government monopoly.

All through the late nineteenth and early twentieth centuries in the United States, efforts were made to nationalize the telegraph carriers, as in Europe. In 1872 Postmaster General John Wanamaker, alarmed at the competition from Western Union, pled for nationalization of the telegraph system: "If the effects of rivalry between the telegraph and the mail upon the revenues of the post-office have not been serious, it is due alone to the liberal management of the latter as compared with that of the companies, a management which, since the invention of the telegraph, has reduced the rates of postage from twenty-five to three cents, and increased ten-fold the correspondence of the country. The natural policy of private companies is to extend facilities slowly and only to profitable points . . . and to reap large profits from a small number of messages, while a Government system . . . pursues exactly the opposite course. Had the policy of the post-office been adopted by the telegraph companies, or had the

Government held to the old rates of postage, the telegraph, instead of now transmitting one-fiftieth part of the annual correspondence of the country, (collecting therefor one-third of the entire expense of the post-office establishment,) would probably transmit at least one-tenth. The profits required of private enterprises would not have permitted such a course. But improvements in telegraphy render it by no means certain that in future the telegraph will not to a very great extent supersede the mail as a means of correspondence."[65]

The improvements that Wanamaker referred to as likely to make telegraphy supersede the mail included multiplexing and facsimile. The telephone, which turned out to be the most important one, was still four years in the future. Because of the rapid adoption of the telephone, telegraphy's growth at the expense of the mails was not the threat that Wanamaker feared; that was reserved for the telephone. By 1890 the revenue of the post office was about $60 million and that of Western Union about $20 million, or the same one-third reported eighteen years earlier.

In any case, the Congress never favored the repeated proposals from the post office for a postal telegraph system. It repeatedly rejected that idea, preferring to require common carrier behavior by private telecommunications carriers. Nondiscrimination was the most important obligation placed upon them. Under common carrier principles telegraph companies were required to treat their competition exactly as they treated themselves. Local telegraph companies were both sources of traffic for each other and also competitors. The Post Roads Act of 1866 specified that to be eligible for its privileges, the telegraph companies had to interconnect, that is, to take traffic from each other. However, imagine two companies, x and y, serving cities A,B,C, and D:

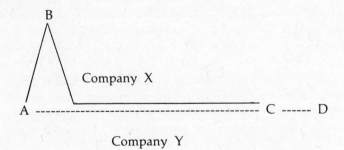

Company y would not happily accept traffic from company x at point C routed from city A to city D. It would prefer to get the revenue for the entire route from A to D. But it would happily interconnect with company x for traffic to or from point B, which it could not otherwise reach. It might even offer special rates from B to C via A. Thus the temptation is strong to treat competing lines less than equitably or, conversely, to offer incentives to use one's own facilities and complementary lines. All this is prohibited: "A telegraph company represents the public when applying to another telegraph company for service, and no discrimination can be made by either against the other, but each must render to the other the same services it renders to the rest of the community under the same conditions."[66]

The reasoning used by the courts and Congress to justify such requirements on communications carriers contains no reference to the First Amendment. The rules are the same as those applied to railroads. None of the arguments rest upon the specially free status of speech in a democratic society. Telegraph customers who went to court claimed denial of rights like those of shipping customers, namely to good commercial service on equal terms to all comers. There were virtually no cases in which First Amendment rights were claimed or such precedents applied.

The main reason that the courts did not see a telegram as an instance of the printed word, and hence equally deserving of First Amendment protection with a printing press, was expense. No one used telegrams initially for debate and self-expression. Although one might send a telegram saying, "Buy 300 at 60 cents max," one would not rush to the telegraph key to vent feelings on a public issue of the day. But that situation was temporary; the technology was changing. With the introduction of private lines and multiplexing, text transmission costs for large users fell by an order of magnitude. The telegraph wires then became a vital vehicle for the press.

Today, costs for electronic transmission of text are falling by three further orders of magnitude. Data transmission costs are falling to levels close to those of the media of mass communication (see graph). In the past a wide gap separated the costs of point-to-point message systems and of mass media. Now low-cost data transmission, the outgrowth of expensive telegraphy, is filling that gap. And as this happens, electronic messages cease to be simply curt aids to transactions; they come to be used for news, education, ideas, and

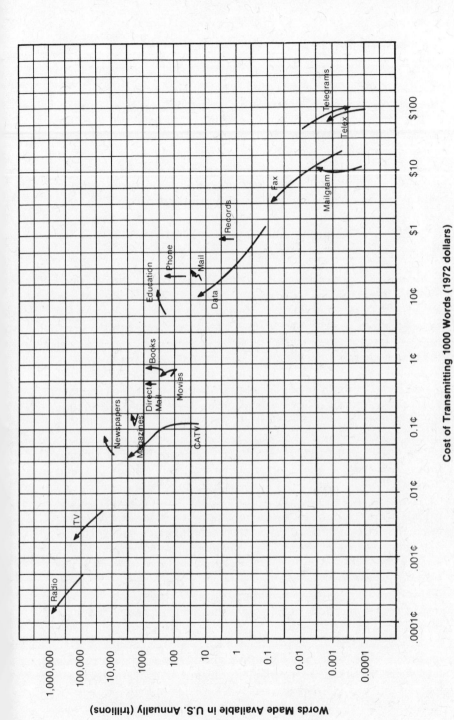

Cost of Transmitting 1000 Words (1972 dollars)

Words Made Available in U.S. Annually (trillions)

Trends in Volume and Cost of Communication by Media, 1960–1978

discussion. Among scientists and hobbyists electronic mail networks already exist on which the active discussion of ideas takes place.

If ten or twenty years hence, homes and offices generally have terminals with messages shown on them at costs less than that of a postage stamp, will the courts still treat telegraphy not as a means of expression but as a transport business subject to regulation as commerce? Or will the courts then recognize electronic expression as an activity about which Congress may make no law?

The Law of the Telephone

Courts like to treat new phenomena by analogy to old ones. When the telephone was invented, the question was whether, at law, the telephone was a new kind of telegraph or something different. If the phone was a telegraph, a body of telegraph law already existed that would apply. The decisions sometimes went one way, sometimes the other; but the model of the telegraph was always there to be considered.

In Great Britain in 1880 the issue of whether at law the telephone was a telegraph was resolved in the affirmative.[67] This was promptly noted in American law journals, and in 1887 similar court decisions calling the telephone a new kind of telegraph started to appear.[68] In one important case in 1899, however, the Supreme Court declined to follow this analogy. The Court held that the 1866 Post Roads Act allowing telegraph companies the free use of timber and a right of way on public lands and post roads did not apply to the telephone, because at the time of the act's adoption nothing was "distinctly known of any device by which articulate speech could be electrically transmitted or received between different points more or less distant from each other."[69] The Court maintained that one of the motives for subsidizing telegraph companies had been to help create a network that the government itself would use. But, said the Court, "governmental communications to all distant points are almost all, if not all, in writing. The useful Government privileges which formed an important element in the legislation would be entirely inapplicable to telephone lines, by which oral communications only are transmitted." This decision deprived phone companies of some useful privileges, but the reverse British decision, by extending the public monopoly status of the telegraph to the telephone, did the private phone companies still more harm, ending in their nationalization in 1912.

Although the general impulse of American courts was to pretend that the phone was nothing new and to take precedents from telegraphy and common carrier law over into the telephone field, some new legal issues appeared with the telephone. One was the question of the evidentiary standing of telephone identification when one could surmise who was at the other end only by voice and not by sight. The question was whether John Doe's statement that Richard Roe had made an agreement during a telephone conversation was admissible in court. While oral agreements had standing in the law, Doe, in a telephone conversation, might not be sure it was Roe who was talking. At first, acceptance of such testimony was found to be subversive of recognized rules of evidence.[70] Eventually, however, the courts adapted to the reality of the vast increase of telephone use in business, and perhaps also to the improved fidelity in telephone reception, by upholding telephoned agreements.[71]

Other legal questions about the telephone concerned the responsibilities and liabilities of a common carrier. The telephone company had to serve all who wanted service, but at issue was how far its liability went in assuring a working connection everywhere and at all times. A basic principle had already been established in telegraph law that the carrier, if not grossly negligent, was liable only for the value of the undelivered physical message, not for the value of the information contained in it. In contrast to transportation common carriers, who could be held liable for the value of the lost or damaged cargo, the message carrier moved something that had no obvious or intrinsic value that the carrier could recognize. The carrier could not be expected to assume a potentially unlimited liability for whatever value some information might have to its users in unknown circumstances.[72] Despite this telegraphic precedent, numerous cases making such claims were brought against telephone companies. A child died without benefit of a physician when the phone failed to work; businesses lost money when their customers could not reach them or when they were mislisted in the phone book. Between 1881 and 1902 the courts of twenty-five states made rulings removing telephone companies from possible suits under the mental anguish doctrine.[73]

The most important legal issue to come up was that of intercompany licensing, specifically the refusal of licenses to competing tele-

phone companies. In the early years of telephony, many cities had
more than one telephone company. In the United States this situation peaked around 1902, when 451 out of the 1002 cities with phone
service had two or more companies providing it.[74] Wherever that
happened, it created a bothersome and unstable situation. Frequently, the person one sought to phone was served by the other
company. Business offices and heavy phone users had to have two
telephones. Generally, the greatest value from having a phone was
obtained by subscribing to the larger of the two systems, for in that
way one could reach more people. As a result, whenever one company became much larger than another, the swing to the bigger system would accelerate, and soon the smaller company would have to
throw in the sponge. The only way for some companies to keep
going for a while was to cut rates, but in the end this meant that they
did not have the capital for expansion. At least within single communities the phone system was a natural monopoly. But this conclusion
was not immediately obvious in the United States. Ideology and the
legal tradition were against monopoly. The benefit to the public of
having telephonic competition was a live issue.[75]

The question about telephone monopoly and competition got into
the courts as a result of adoption of regulations by cities, states, and
the federal government requiring a prospective phone company to
obtain a license. Before the license could be issued, the utility had to
satisfy the licensing authority that its service would be "in the public
convenience." For example, the state of Ohio required a telephone
company to secure a certificate of convenience from the Public Utilities Commission before exercising a franchise in any town where
there was already a telephone company furnishing adequate service.[76] In the village of Mendon, where one company was already
providing phone service, another company obtained a franchise to
do the same but was refused a certificate of necessity by the state on
the grounds that it was not in the public convenience to have two
telephone companies. The Supreme Court of Ohio in 1921 sustained
the denial.

The court reasoned entirely in terms of common carrier law,
making no reference to freedom of speech. It concluded that both
companies were common carriers and that in the case of a utility the
government could decide whether the public interest was better
served by having competition or not: "It is important to notice that
the section does not prohibit another company from competing, but

makes it a condition precedent . . . to first apply for and receive a certificate from the Public Utilities Commission. The Commission in the act is provided with all the facilities to investigate and determine whether the public convenience will be served . . . It is thus clear that the Commission has the authority to deny the right, and this authority is dependent upon and turns upon public convenience and public welfare."[77]

This conclusion is unarguable, unless the facility that the government proposes to regulate is one that is excluded from the domain of public regulation. Under the First Amendment the reasoning of the court could not have been applied to printing presses. Deciding whether there should be one press or more, and who should run them, is clearly excluded from political control. Why not telephone systems? Are they not equally obviously instruments of speech? The extraordinary fact is that this question did not occur to either the Ohio court or to any other court that dealt with the licensing issue, or even to any of the lawyers arguing the cases. The issue simply did not arise. The telephone was seen as a successor to the telegraph; the telegraph in turn was seen as a common carrier like the railroad; and so that was the law applied. The phone was not seen as a successor to the printing press.

Furthermore, the Ohio court, in dealing with constitutional guarantees, focused on those in the Fourteenth Amendment, not the First. The constitutional issue raised by the case was not freedom of speech or of the press but the right of property. The issue addressed by the court was whether it was unconstitutional under the due process clause to deprive the telephone company of its property. The court's conclusion came straight out of public utility law. It relied on the classic case, Munn v. Illinois: "When private property is devoted to a public use, it is subject to public regulation."[78]

The same kind of partial reasoning appears in Supreme Court decisions about the licensing of telecommunications. The 1934 Communications Act requiring the FCC to make a finding of "public convenience, interest, or necessity" before issuing a license to a radio common carrier was tested in 1952 in FCC v. RCA Communications, Inc., a radio telegraphy case that became the precedent for telephony.[79] The issue in the case was that the FCC had licensed the Mackay Radio and Telegraph Company to open new radiotelegraph circuits to Portugal and the Netherlands in competition with circuits already established by RCA. The FCC's finding that these were in the public interest was based entirely on the allegation that, as there was a "national policy in favor of competition," competition should be allowed wherever "reasonably feasible." The Court, in a decision by Frankfurter, the least likely of the recent great figures on the Court to see any free speech implications, disagreed: "Surely it cannot be said in these situations that competition is of itself a national policy. To do so would disregard not only those areas of economic activity so long committed to government monopoly as no longer to be thought open to competition, such as the post office . . . and those areas, loosely spoken of as natural monopolies or—more broadly— public utilities, in which active regulation has been found necessary to compensate for the inability of competition to provide adequate regulation."

The court decision chose not to interpret the congressional criterion of "public convenience, interest, or necessity" as a vague and expansive phrase in effect telling the FCC it could issue licenses whenever any useful purpose was served and without having to meet special tests. Instead it interpreted the phrase restrictively as imposing on the FCC the requirement to test every license against some finding as to why the issue of the license was positively desirable. The phrase "public convenience, interest, or necessity" can be read either way, particularly as the conjunction used is "or," not "and." Convenience to the public, manifested by the confidence of investors that customers are there, might be basis enough for granting a license. This broad, expansive phrase could easily be interpreted as Congress telling the FCC to issue licenses generously in terms of any kind of showing that some benefit is to be served. But such was not the way that regulatory agencies in the past, or the courts in reviewing them, chose to interpret it. They chose, as did the Supreme Court here, to interpret it as a mandate to the government to regulate: "To say that national policy without more suffices for authorization of a competing carrier wherever competition is reasonably feasible would authorize the Commission to abdicate . . . one of the primary duties imposed on it by Congress . . . the Commission must at least warrant . . . that competition would serve some beneficial purpose such as maintaining good service and improving it."

This view turned the whole set of presumptions under the First Amendment on its head. Even under the most eviscerated interpretation of the First Amendment, the presumption is made that anyone can engage in communicative activities freely, unless there are overbalancing considerations. One such balancing consideration might be the inconvenience of financially weak and noninterconnected common carriers. But in discussing the licensing of competing telecommunications companies, Frankfurter saw no need to balance other considerations against a presumption of freedom, as even he would have done if he had thought the First Amendment relevant. The presumption was exactly the opposite: no license to communicate was to be issued unless the state in its majesty concluded that such was desirable.

Perhaps the Court could have reached the same decision by distinguishing the technology of telecommunications networks from that of printing presses. Interconnectedness is important for the former and not for the latter. But no such argument was made. The Court did not even recognize that the issues being dealt with related to the precedents of First Amendment law. This silence, more than the finding itself, was extraordinary. In decisions about common carriers the First Amendment had simply disappeared. Even the two First Amendment absolutists, Black and Douglas, each of whom wrote a dissenting opinion but on opposite sides, failed to mention the First Amendment. Black thought that the FCC had produced enough evidence on behalf of its decision. Justice Douglas, who on a recognized First Amendment case would never have allowed a government agency to exercise its judgment of the public benefit, wrote: "I agree with the Court that it is necessary under the Federal Communications Act to establish that the licensing of a competitive service offers a reasonable expectation of some beneficial effect . . . But . . . existing facilities are in excess of those required to handle present and expected traffic; the proposed operations will redistribute present traffic rather than generate new traffic."

The RCA case became the precedent for telephony in 1974 in a case in which RCA was on the other side.[80] RCA had been licensed by the FCC to offer leased-line voice service via Comsat in competition with the Hawaiian Telephone Company. The District of Columbia Court of Appeals once more ruled that the FCC had not done its homework in making a positive finding of the desirability of having this second channel for speech: "The FCC has not conformed to the requirement that it find the public convenience and necessity dictate the new service . . . When the FCC considers an application for certification of a new line, it must start from the situation as it then exists, and must apply the statutory standard to determine whether indeed the public convenience and necessity require more or better service."

The burden of proof has here been shifted. It is no longer enough for the FCC to say that it has considered the matter and has concluded that the public convenience or public interest would be served by having competing services. Now the court asks if public convenience and necessity "dictate" a new service, or whether public convenience and necessity "require" more or better service. A prohibition on the establishment of instruments of communication unless the government concludes they are required is all that the most authoritarian regime could want as public policy.

First Amendment considerations have been disregarded in telephone law in a number of respects other than restrictive licensing. For example, the Supreme Court, which has deemed special taxes on newspapers to be unconstitutional, has no such problems with taxes on phone bills.[81] Also, denial of a telephone to a subscriber for having used bad language was sustained as far back as 1885.[82] Yet denial of a printing press to a publisher because of the obscenity of previous publications would be unconstitutional prior restraint. One can wonder, therefore, whether the prohibition on licensing and prior restraint that is so central to America's system of freedom will survive when most communication comes to be transmitted by electronic carriers rather than in meetings and the press.

At the same time, though common carrier doctrine often lacks explicit reference to civil liberties, many of the same concerns are dealt with in different words. In its own way the law of common carriage protects ordinary citizens in their right to communicate. The traditional law of a free press rests on the assumption that paper, ink, and presses are in sufficient abundance that, if government simply keeps hands off, people will be able to express themselves freely. The law of common carriage rests on the opposite assumption that, in the absence of regulation, the carrier will have enough monopoly power to deny citizens the right to communicate. The rules against discrimination are designed to ensure access to the means of communication in situations where these means, unlike the printing press, consist of a single monopolistic network. Though First Amendment precedents are largely disregarded in common carrier law, still this one element of civil liberty is central to that law.

In one important respect, however, this libertarian element of common carrier law has eroded in the electrical era. When the issues of common carriage had to do with the post office, there was substantial awareness of the carrier's relevance to the political life of the country. When new electrical carriers came along, that awareness disappeared. The new carriers seemed unrelated to public affairs. The courts treated them simply as instruments of commerce subject to any regulation the government chose to impose.

6. Broadcasting and the First Amendment

Two systems, publishing and common carriage, each in its own way seeks freedom in communication. The first minimizes public controls, allowing anyone to establish facilities for expression. The second obliges the carrier to serve all customers equally. Each system has a logic for its circumstances, which are an open market on the one hand and a monopoly on the other. A very different and much less coherent system has evolved for broadcasting. In broadcasting, freedom of speech and of the press has been compromised. This is true in America and even more so elsewhere. Full, robust citizen participation in a democratic forum casts only a shadow on the tube.

The manner in which Congress and the courts moved away from classical First Amendment concepts differs fundamentally for radio broadcasting and common carriers. When the legal system for the electrical carriers, telegraph and telephone, was being created, policy makers were to a large extent unaware that speech and press were being regulated by their decisions. When in the 1920s the broadcasting system was being created, they were very aware. There was no inadvertence about the relevance of the First Amendment in the discussions of 1927 that led to a radio act. The applicability to radio of the free press system was widely, if not foresightedly, discussed. When Congress chose to regulate radio speech rather than to conform literally to the constitutional injunction that it should make "no law abridging freedom of speech," it did so not because that clause seemed irrelevant, but rather because it seemed impractical when applied in the new technological context.

The options at the time for the organization of radio broadcasting were several. It could have been set up as a government monopoly, a common carrier, a regulated commercial activity, or an unregulated market activity like publishing. Of these options, only the last was not considered. In America the government monopoly and common carriage options were consciously rejected in favor of a regulated system.

A Rejected Option: Nationalization

In the early 1920s in Europe, socialist ideology was at its historical peak. Social Democratic parties had become major parties in Germany, France, Great Britain, and elsewhere. Faith in the benevolence of democratically chosen political authorities was strong. The remedy for the evils of self-interest and cupidity seemed to be to place a greater range of activities under social control. Furthermore, belief in the rationality and efficacy of national planning was shared not only by the left but also by large segments of the center and right. The idea of turning a new and invaluable resource where vested interests had not yet been created over to the very private enterprise system that the socialist movement was trying so hard to restrict, if not to abolish, was offensive to large wings of European public opinion.

But another force was even more effective than ideology in the decision to nationalize broadcasting. This was the bureaucratic impulse of both national security organizations and post, telegraph, and telephone authorities who wished to keep radio in their hands. These antisocialists, as most of them were, shared with the left a belief in radio nationalization, and they probably understood the character of this proposal far better than did those socialist idealists who provided its rhetoric. The result in some countries, like France, was direct national management of broadcasting, and in other countries, where liberal concerns for freedom from state domination were more acute, the result was the creation of an independent broadcasting authority, such as the British Broadcasting Company (BBC).

In Britain the post office, like AT&T in the United States, from the beginning saw radio as a natural extension of its activities in telecommunications. Transmission through the air was simply a new method for sending messages. The chief engineer of the British Post Office, Sir William Preece, had done pioneer research in wireless telegraphy. From 1892 he experimented with transmission by electrical induction across Loch Ness and the Skerriel, and by 1899 his signal spanned three miles. When Marconi and others began experimenting with the use of Hertzian waves instead of induction, the post office was interested, and Preece played a key role in introducing Marconi to those who mattered in England.

During the first two decades of the twentieth century, when radio transmission became a proven reality, numerous amateurs experimented with and helped develop the art. To exercise their equipment, they needed to communicate with each other. They talked, played music, and discussed the improvements that they had made over the air, all the while asking when, how far, and how well reception was coming through. The Marconi Company did the same, even putting Lauritz Melchior and Dame Nellie Melba on the air.[1]

The British Post Office recognized this not only as a useful and necessary activity but also as a threat to what they and the armed forces considered to be a proper and organized use of the medium. Their concern was expressed when in 1920 an entertainment program interfered with radio communication with an airplane lost in fog over the English Channel. Restrictive regulations were imposed. After 1904, all wireless transmitters had to be licensed by the post office. The license given the Marconi Company in 1920 required separate approval for each broadcast on the theory that each was an experiment in transmission that could interfere with radio telegraphy. At one point the broadcasts were stopped by the post office. Amateurs with ten-watt transmitters were limited to two hours a day, and then only if the post office was convinced that they had a serious scientific purpose. At first no music was allowed. By 1921 the post office had issued 150 amateur transmitting licenses and 4000 receiving licenses. The regulations were widely violated by amateurs, who repeatedly petitioned that "every Englishman is entitled to hear what is going on in his aether."[2]

Annoying as the amateurs were to the post office and the defense establishment, fear of commercialism was what finally led to the BBC. Once the British public heard about American popular broadcasting, there was no banning it completely; it could at most be restricted. The broadcasting scene in America with all its confusion filled British observers with horror. A characteristic attitude was expressed by one of Britain's radio pioneers, R. N. Vyvyan: "[In the United States] newspapers and big retail stores soon saw that broadcasting offered a wonderful opportunity for advertising their wares . . . There were no regulations which forbade this . . . It did not matter whether one station were interfering with another . . . as there was a boom in broadcasting, and everyone was going to get in on the aether. By the middle of 1923 there were over 500 broadcasting stations in America and an audience of about 2 million. By 1924 the number of stations had grown to 1105, and the 89 wavelengths available had to be shared by them; chaos in the aether was of course the result. Great Britain was slow to start on general broadcasting. The restrictions imposed on private activities in England, while frequently annoying, are generally for the public good, and in wireless matters the British Post Office invariably investigate a new proposal thoroughly before sanctioning it . . . American practice showed . . . the necessity of confining the transmission of programmes to one authority . . . In 1921, after many discussions, the Post Office authorized a very limited broadcasting to be carried out from the Marconi Company research station at Writtle . . . Other organizations besides the Marconi Company were interested in broadcasting and were permitted to build stations. Conflicting interests eventually brought about an agreement to form a British Broadcasting Company, financed by these various interests, but under severe restrictions, wisely imposed by the Post Office. The British Broadcasting Company was not allowed to use the stations for advertisement."[3]

In the United States, none of the conditions prevailed that led to the nationalization of broadcasting in Europe. Socialism was weak: Eugene Debs won an all-time high of 3½ percent of the presidential vote in 1920, and at no time was there more than one Socialist Party representative in Congress. Telephony and telegraphy were private enterprises; the post office had no part in electrical communication. There was general antipathy to government ownership. Although during World War I amateurs were shut down and the government took over all radio transmission, at the end of the war when bills were introduced in Congress to perpetuate the government monopoly, the amateurs and the industry launched a successful protest.

Support for the European scheme of nationalized broadcasting organization was rare in the United States. The navy was one of its main advocates. In 1918, Representative Joshua Alexander unsuccessfully introduced a bill to nationalize all radio transmitters and turn over control of their use and sublicensing to the navy. A navy broadcasting monopoly, argued Admiral W. L. Rodgers in 1924, could best provide global transmission of news at low cost: "How can radio be better utilized so as to give news of public interest without such exclusive reliance on the press?"[4]

Strange bedfellows for the navy came from the left. Bruce Bliven, the reform journalist who later became publisher of the *New Republic*, in 1924 considered a nationalized system: "Radio broadcasting should be declared a public utility under strict regulation by the Federal authorities; and it may be necessary to have the Government condemn and buy the whole industry, operating it either nationally or locally on the analogy of the post office and the public school system."[5]

But far more typical than this social-democratic idealism was the American perception of the BBC and similar foreign public monopolies as censorious organizations that were holding back broadcasting development compared to its explosive growth in the United States. One critic in 1936 likened the policy of New York's municipally owned station WNYC to that of the BBC which, to avoid inciting the Postmaster General to censorious action, "reduces political controversy and political education on the air to an absolute minimum. It brings about a strong ambition to seek neutrality and colorlessness."[6]

The alternative of a government monopoly was never seriously considered in America. Government regulation as a utility affected by the public interest was as far as the leaders of opinion would go, and such regulation was viewed as an unfortunate necessity to be minimized. But the United States was the exception. In most countries the authorities accepted only grudgingly the use of scarce radio frequencies for frivolous use in entertainment, and they restricted such broadcasting to a "responsible" monopoly, consisting either of government officials or specially authorized public bodies.

The American Choice: Regulation

Before the beginning of broadcasting, point-to-point communications users and amateurs struggled over their rights to the airwaves. By 1919 about four thousand amateur stations existed, four or five times the total of commercial and military stations.[7] Aware of the struggle that amateurs had had to wage against the navy in order to retain access to the airwaves, the early broadcasters were already sensitive about their right to be heard. The first broadcast is generally dated as election night 1920, when KDKA in Pittsburgh went on the air with the returns. In the period 1920–1923 attention focused largely on the physical efforts to get started. But with the increase in transmitters, interference became a major problem.

So, from 1924 till passage of the Radio Act of 1927, Congress and the press debated over how to structure the American broadcasting system. Selective licensing and free speech became major issues. Following a brief hiatus in policy debate as observers waited to see how the act would work out, controversy intensified again from 1930 to 1934. The gradual tightening of federal regulation evoked sharp criticism of censorship. Those who had been concerned with the censorious implications of the 1927 act had reason to believe that they had been proven right. Nonetheless, in 1934 Congress passed a new Communications Act, which also dealt with common carriers but for radio reenacted most of the provisions of the act of 1927.

The Technological Context: Spectrum Shortage

By 1925, the available frequencies for broadcasting seemed to have been exhausted. There was general consensus that room could be made for no more broadcasting channels than the 89 that had been allocated. "Conditions absolutely preclude," Secretary of Commerce Herbert Hoover warned, "increasing the total number of stations in congested areas. It is a condition, not an emotion."[8]

The licensed broadcasters agreed. Already ensconced in their frequencies, they favored keeping additional competitors off the air. The National Association of Broadcasters in 1925 recommended that no more licenses be issued because the saturation point had been reached.[9] This was a pleading by special interests, but it was also a consensus belief. Morris Ernst, a civil liberties lawyer, conceded the reality of saturation and the consequent inevitability of government choice among applicants. Without a selective process, he advised, the airwaves would be like Forty-Second Street at rush hour.[10]

Yet the law did not permit the Secretary of Commerce to make a selection of licensees. The Radio Act of 1912, which was designed to ensure that amateurs did not interfere with the navy, gave the Secretary authority to issue licenses for transmission, but this power was only that of a traffic policeman, assigning frequencies so as to minimize interference. The Secretary had no authority to refuse a license.[11]

Although mandated to issue licenses to all applicants, the Secretary tended to give the best frequencies to those broadcasters he favored. Special interest applicants like religious or labor groups were given part-time licenses on inferior channels; commercial broadcasters were given better ones.[12] In 1926 even this limited power to regulate frequencies and hours of operation was lost to the Secretary. The Second National Radio Conference in 1923 had opined that the Secretary had that power, but when Secretary Hoover sought to penalize the Zenith Radio Corporation for operating on an unauthorized frequency, a court held that the 1912 Radio Act did not permit enforcement.[13]

Broadcasters all over the country promptly started jumping to desirable frequencies. Interference became rampant. This crisis triggered action in Congress, which had been discussing a new radio act for years. Legislation providing for selective licensing of broadcasters in the public interest seemed inescapable.

In retrospect, selective licensing of favored applicants was not unavoidable. Alternatives were available, and even under the scheme of regulated utilities the pressure of spectrum shortage was not so great as it seemed at the time. One measure that was taken to reduce the need for the authorities to select among license applicants was to force broadcasters to share frequencies. Extensive sharing, however, would have given each licensee access to the air for only a few hours a week. As a general solution, such sharing was rejected on economic grounds. As the art of broadcasting progressed, the capital cost for a good station rose to more than $150,000, with operating costs of $100,000 or more per year. Hoover observed that the costs "are in large part the same whether the station works one day in a week or seven."[14] So the multiplication of stations with each allowed only limited time for broadcasting would have driven down the earnings and therefore the quality of broadcasts by the stations. This objection could have been met by stations sharing both frequencies and the costly transmitter plant, but such a variant of common carriage was rejected by congressional opinion.

Congress failed to recognize the possible transiency of spectrum scarcity. By the mid-1920s, there was awareness that the progress of technology might eventually overcome the shortage, but that seemed only a vague possibility. The political decision makers were neither aware of what existed in laboratories nor willing to consider using expensive technology for multiplying channels when some channels, though a smaller number, could be had cheaply. Amateurs had demonstrated that frequencies above those of the standard broadcast band could be effectively put to use. They had been limited by the 1912 Radio Act to frequencies higher than 1500 kHz, widely regarded as useless, in hopes that they would fade away. Instead, in 1923 and 1924 they achieved transmissions at 3 MHz, the bottom limit of what later came to be called high frequency (HF) or short wave, across the Atlantic and from Australia to California. In 1924 the British Marconi Company ran experiments at frequencies of 10

MHz, and by 1926 it started a telegraphic service for Canada at 10–20 MHz. In 1925 it was studying frequencies of 150 MHz, now called VHF.[15] Today, as a result of the expansion of usable frequencies, only about 2 percent of that spectrum is now devoted to broadcasting for both radio and television, but of the frequencies that policy makers considered in the 1920s, as much as half was dedicated to radio broadcasting. They did not foresee that dramatic change.

Some news of the prospective availability of additional frequencies did leak through to the political authorities. As early as 1924 Hoover urged passage of a temporary new radio act until there was time to study the implications of new developments in the radio art.[16] In 1926 when Stephen B. Davis of the Department of Commerce was asked in a Senate hearing to name the shortest wavelength that was practicable to use, he replied that until recently short waves had been considered almost worthless, and that despite promising experiments, no one yet knew what was going to become of them.[17]

When in 1927 Hoover spelled out his reservations about extension of the usable spectrum for broadcasting, they were economic and political rather than strictly technical: "It has been suggested that the remedy lies in widening the broadcasting band, thus permitting more channels and making it possible to provide for more stations. The vast majority of receiving sets in the country will not cover a wider band, nor could we extend it without invading the field assigned to the amateurs."

As stations raised their power to 500 watts to improve their signal, interference became more widespread. Yet even the "great engineer," Hoover, saw no short-term prospects of a technical fix that would provide for strong stations with good service while also meeting the demand for new stations, which had reached the figure of 250 applications awaiting action: "It is a simple physical fact that we have no more channels. It is not possible to furnish them under the present state of technical development . . . All these things bring us face to face with the problem which we have all along dreaded [selection of licensees] and for which we have hoped the development of the art might give us a solution; but that appears to be far off, and we must now decide the issue of whether we shall have more stations in conflicting localities until new discoveries in the art solve the problem."[18] Senator Hugo Black, who would later as a member of the Supreme Court become the spokesman for First Amendment absolutism, as late as 1929, in proposing cross-ownership rules against broadcast licenses for public utilities, looked toward a future in which technology might eliminate the need for such constraints— "the possibilities of radio cannot be foretold today"—but saw for the moment no alternative except to choose desirable broadcasters.[19]

There was thus a vague awareness of possible future technological solutions to spectrum shortage, but it was not informed by knowledge of the state of the art. There was a consensus on the immediate necessity of licensing, but in the minds of many policy makers this conclusion coexisted in deep cognitive dissonance with a desire to maintain freedom from government control for the new medium.

The Regulatory System

In the days of Milton, licensing was imposed on publishing by the Crown with a view to restricting the press. In the United States, on the contrary, the intent in imposing licensing was to promote radio expansion, though without clear understanding of the economic and technical implications of what was being done. Gradually and with reluctance both the government and the industry moved in the 1920s toward a regulated regime.

All the while everyone was disclaiming censorship. Hoover warned that monopoly control of radio would be "in principle the same as though the entire press of the country was so controlled." He opposed "the Government undertaking any censorship even with the present limited number of stations."[20] He nevertheless suggested in 1924 that the Commerce Department be given jurisdiction over the power, wavelength assignment, and broadcast hours of radio stations. The National Association of Broadcasters protested that the Secretary would thereby obtain "Napoleonic Powers" and instead proposed regulation by a communications commission.[21] With crowded airwaves becoming a serious problem, some type of regulation seemed inevitable; the debate was over its form. Broadcasters adopted resolutions against censorship of radio programs and limitation on advertising.[22] Secretary Hoover proposed a plan involving federal allocation of wavelengths and amounts of power which would leave "to each community a large voice in determining who are to occupy the wave lengths assigned to that community."[23]

The Fourth National Radio Conference in 1925, which in effect spoke for the industry, agreed that the federal government "must have the right, through issue of licenses, control of power, assignment of wavelengths, and other appropriate measures, to handle as a whole the interstate and international situation." But government authority "should not be extended to mere matters of station management not affecting service or creating interference, nor should it under any circumstances enter the forbidden field of censorship." They recommended that Congress stipulate that "the administration of radio legislation shall be vested in the Secretary of Commerce," that "the doctrine of free speech be held inviolate," and that "those engaged in radio broadcasting shall not be required to devote their property to public use and their properties are therefore not public utilities in fact or in law." Broadcasters then, just like cablecasters today, were pleading not to be compelled to assume the responsibilities of common carriers. The industry further resolved that a radio license "shall be issued only to those who, in the opinion of the Secretary of Commerce will render a benefit to the public, or are necessary in the public interest, or are contributing to the development of the art," and that "no monopoly in radio communication shall be permitted." This last was an attack on AT&T's common carrier plan.[24]

As the number of stations grew and interference worsened, the pressure for regulation grew. In the design of the regulatory scheme, among the most debated issues was censorship. In 1926 Representative Wallace White stated that "sooner or later there will have to be some [governmental] authority to establish priorities as to subject matter," but he dropped reference to this subject from his proposed radio bill because of people's fears that such a provision would confer censorship powers. He assured Representative Fiorello LaGuardia that the bill gave the Secretary of Commerce "no power of interfering with freedom of speech in any degree."[25] The Commerce Department reaffirmed the government's unwillingness to undertake censorship.[26] But others were less categorical against a role for government. Some argued that profit-making radio stations should be regulated as public utilities; others pointed to the difficulty of deciding what is in the public interest.[27] A proposed prohibition of any radio broadcasts concerning the theory of evolution was rejected by Congress.

A split developed, partly along party lines, between proponents of station licensing as to whether it should be by the Secretary of Commerce or by an independent commission. Democrats recognized in Hoover a Republican presidential contender. The Senate Democratic leader, Joseph T. Robinson, charged that executive branch control of radio would deprive Democrats of radio time: "It is in conflict with recent and more ancient experience to entertain that freedom of opinion and of speech will be permitted by giving the President power to dominate a principal agency of publicity."[28] President Calvin Coolidge, on the contrary, argued that there were already enough government agencies and so radio regulation should remain in the Commerce Department. Partly in response to difficulties encountered by two senators in broadcasting their views, the Senate 1926 bill draft included an independent commission and a prohibition against stations broadcasting any material that discriminated in favor of candidates for political office or discriminated in the discussion of any subject affecting the "public interest," a formula which foreshadowed the present fairness and equal time doctrines.[29]

However, given Congress' ambivalence and conflicting views, it was unable to reach agreement on a bill until the Zenith case brought chaos to the airwaves. In July 1926 the Attorney General, pursuant to the court decision in that case, issued an opinion requiring the Secretary of Commerce to grant a license to any applicant, leaving him without power to specify wavelengths.[30] As rampant interference disrupted reception, the *New York Times* editorialized that broadcasting must be regulated, and the Department of Commerce should do it, for radio's "manifest destiny" was, as in England, to be a controlled monopoly.[31] The country's leading newspaper did not perceive an issue of free speech.

Spurred to action, on February 23, 1927, Congress passed a new Radio Act, which created the Federal Radio Commission as the regulatory authority. But even adherents of the new law were ill at ease. Senator David I. Walsh complained that the bill "fails to clearly and definitely safeguard the rights of free speech to prevent the control of broadcasting in the interest of the dominant party or powerful special interests, and to secure to the exponents of all shades of opinion a reasonable access, upon equal terms to its facilities for influencing public opinion."[32] The bill contained four sections which had a bearing on censorship. Section 18 was an equal time provision which required stations, if they chose to air views of political candidates, to treat rival candidates equally, and which gave stations "no power of censorship over the material broadcast under the provisions of this paragraph." Section 29 provided that nothing in the law "shall be understood or construed to give the licensing authority the power of censorship over the radio communications or signals transmitted by any radio station, and no regulation or condition shall

be promulgated or fixed by the licensing authority which shall interfere with the right of free speech by means of radio communications.
No person within the jurisdiction of the United States shall utter any
obscene, indecent, or profane language by means of radio communication." Section 11 set the stage for years of controversy in its requirement that the Radio Commission issue a license if it determined
that "public convenience, interest, or necessity would be served by
the granting thereof." And Section 13 forbade the commission from
licensing any person or corporation which had been found guilty of
monopolizing or attempting to monopolize radio communication.[33]

These provisions remain even today the law of the land. In substance, the system of 1927 is the system still. What emerged in the
law was a balancing act that bothered many legislators: an arrangement under which broadcasters were chosen by the government and
licensed, but at the same time the licensing authorities were enjoined
not to control content or censor the broadcasters.

The Growth of Censorship

In the early 1920s, censorship of speakers by radio stations themselves, sometimes called "private censorship," was common. As
early as 1921 an emergency switch in the studio of a Newark station
was used by the engineer to cut off speakers in mid-sentence if their
material was deemed unfit for public ears.[34] Subjects treated in this
way included birth control, prostitution, and cigarettes.

The advertisers' interest in large audiences made broadcasters
sensitive to "public opinion." It was common for members of the
listening audience to send cards of approval or disapproval to stations; some radio dealers even provided forms for the purpose. Station managers avoided topics that might arouse community ire. In
1923 the *New York Times* commented, "The radio audience is so large
and represents such a varied interest that the censor must eliminate
anything which might injure the sensibilities of those listening."[35]

A common justification for this censorship was the perception of
radio as a uniquely powerful force. Though it was sometimes viewed
as "publishing," it was often looked upon as a potentially more dangerous instrument which could, without vigilance, destroy American
ideals.[36] Secretary of Commerce Hoover in 1924 said: "Radio has
passed from the field of an adventure to that of a public utility. Nor
among the utilities is there one whose activities may yet come more

closely to the life of each and every one of our citizens, nor which holds out greater possibilities of future influence, nor which is of more potential public concern. Here is an agency that has reached deep into the family life. We can protect the home by preventing the entry of printed matter destructive to its ideals but we must double-guard the radio." President Coolidge echoed these sentiments: "In this new instrument of science there is an opportunity for greater license even than in the use of print, for while parents may exclude corrupting literature from the home, radio reaches directly to our children."[37]

In 1924 a conflict arose in California over a radio speech advocating the private ownership of water rights. Partisans of public ownership got the radio inspector to threaten censorship. The radio industry protested this abuse of authority.[38]

While cases of censorship hit both left and right, liberals worried that radio might become "an organ of orthodoxy," dominated by powerful private interests. They turned to government to protect free speech against private censorship.[39] Representative Wallace White proposed a Bureau of Radio within the Department of Commerce to "prescribe the nature of the service to be rendered and the priorities as to subject matter to be observed." The *New York Times* grumbled, "The bill virtually makes the Secretary of Commerce and the board acting under his direction censors of intelligence broadcast by radio."[40]

In a symposium on freedom of the air in the reform magazine *The Nation* in 1924, all the statements opposed censorship, but in it Grover Whalen, a New York City political figure, looked to government to protect the public both from poor broadcasts and from private censorship by station owners: "Broadcasting should be as free as the air through which the sound waves are impelled, except for such government control as may be necessary and advisable." David Sarnoff of RCA countered that radio was becoming a major medium of opinion to which "the same principles that apply to the freedom of the press should be made to apply." He emphasized the danger of government censorship and minimized that of broadcaster controls. With over five hundred stations broadcasting to an estimated daily audience of ten million, there was no danger of anyone monopolizing what went out on "the vast reaches of air." The "real danger," he argued, was "in censorship, in over-regulation." Public opinion must be the test of what is broadcast.[41]

Other industry spokesmen, assembled the same year at the Third

National Radio Conference, likewise opposed government regulation of program material but accepted the premise that radio should stay "clean and fit for home consumption at all times." The broadcasters, they were confident, could ensure this. The industry rule of the day was: censor ourselves so the government will not.[42]

The most controversial incident of private censorship concerned H. V. Kaltenborn, a popular radio personality and associate editor of the *Brooklyn Daily Eagle*, who broadcast on AT&T's New York station WEAF. In 1924 he criticized Secretary of State Charles Evans Hughes for rejecting the Soviet Union's bid for recognition. When Hughes heard the broadcast, he called an AT&T representative, and word was relayed to New York that "this fellow Kaltenborn should not be allowed to criticize a cabinet member over the facilities of the New York Telephone Company." Kaltenborn's contract was canceled.[43]

Kaltenborn later articulated the free speech and common carrier issues as they arise in the context of limited radio channels: "Broadcasting stations need the cooperation of Federal authorities . . . The Department of Commerce allocates and withdraws wavelengths and broadcasting rights. A corporation controlling broadcasting stations . . . would be foolish to prejudice its interests by antagonizing those in high places . . . Unknown to the general public, there is a thoroughgoing radio censorship already in effect. It operates quietly and efficiently through a process of exclusion. Those who are excluded have thus far had the recourse of opening stations of their own, but this will soon be cut off. Federal policy from now on will oppose the erection of an unlimited number of high-powered stations in congested districts. Before long, we are likely to witness a legal battle to compel broadcasting stations which make a practice of selling time, to sell it to all comers on equal terms. 'Freedom of the air' will come to have meaning akin to 'free speech' or 'freedom of the press.' Broadcasting is as much of a public service and convenience as the telephone, and ultimately must be subject to the same kind of regulation and control."[44]

Though the broadcasters defended their exercise of censorship by analogizing it to a newspaper's editorial privilege, their opponents argued that it was not as easy to open a radio station as to start a newspaper and would soon become more difficult, and that "editorial privilege" becomes suspect if it is exercised under permit from government authorities. As an AT&T executive wrote during the Kaltenborn controversy, the company had a "fundamental policy of constant and complete cooperation with every government institution that was concerned with communications," which was hardly in the tradition of the free press.[45]

Instances of private censorship continued.[46] Stations cut off or barred speakers on such diverse grounds as their left-wing politics, opposition to prohibition or to chain stores, mention of birth control, criticism of the government, and, predictably, criticism of the "radio trust." In 1926 a speech by the socialist leader Norman Thomas, scheduled to be broadcast over station WMCA in New York, was abruptly canceled. Thomas protested that no station would risk criticizing the administration so long as licensing was in the hands of a cabinet officer. WMCA relented, and a week later Thomas aired his speech.[47]

The dilemma of free speech on a scarce medium left even ardent civil libertarians in conflict. Ernst, testifying for the American Civil Liberties Union in 1926, conceded the need for radio legislation because, unlike newspapers, "You have got a limited territory to divide," but a mere legislative injunction that free speech be not impaired did not go far enough.[48] A brisk trade, he noted, had already developed in licenses, which were sold for exorbitant sums. The Chicago Federation of Labor, for example, had difficulties obtaining a station since the Commerce Department's position was that the Chicago airwaves were saturated. As a result, this organ of working men faced having to pay $250,000 to buy out someone else's permit. Ernst summed up the problem: "So long as the Department can determine which individuals shall be endowed with larynxes it does not need additional power to determine what shall be said." He nevertheless concluded ambivalently that, "Short of government ownership and control of the stations, some machinery should be set up to insure as far as possible the presence on the air of minority points of view." But on the machinery to do this he waffled between government control and distrust of it: "Granted that some censorship of the air is at this time an engineering necessity, those who believe in the right of free speech must see to it that this censorship is controlled so far as possible by the listening millions of the country. To vest unchecked control in the Secretary of Commerce and a super-political Committee of Five would, in effect, be a partial deeding away of our aerial voices, our ears, and consequently, our very thoughts."[49]

The Radio Act of 1927, with its prohibition of censorship by the Radio Commission, had hardly become law before it was acknowledged that the very power to license involved inherent censorship. Radio Commissioner Henry A. Bellows explained that because of the shortage of available wavelengths, the Radio Commission was forced to make choices: "The physical facts of radio transmission compel what is, in effect, a censorship of the most extraordinary kind."[50] The *Columbia Law Review* made the same point: "The standard of 'public convenience and necessity' seems to afford a sufficiently effective device to guarantee the freedom of the air. What the character of its program is, and, more particularly, whether on controversial questions a station has given fair representation to both sides would easily seem important elements in the determination of whether or not it should be permitted to continue broadcasting."[51]

For about two years following the passage of the Radio Act, while the commission went about its business of granting or denying licenses and trying to flesh out the meaning of "public convenience, interest, or necessity," there was a hiatus in policy debates. Licenses were granted to the National Preparedness Movement and to station WEVD, a socialist outlet in New York dedicated to the memory of Eugene V. Debs, on the grounds that it was not improper for a station to be the "mouthpiece of a political minority." Licenses were refused or made probationary in the case of stations that were used for the broadcasting of personal attacks or disputes, which the commission found disagreeable.[52] Although the commission continued to disavow censorship, it took into consideration such factors as the distribution of different types of service among various stations, avoidance of duplication of programs, and availability of the station's service to the public in another form, such as phonograph records.

The Radio Commission also made bold declarations about what fell within the constitutional guarantee of free speech. Free speech, said the commission, clearly extended to matters of opinion on political and religious questions, but: "Does this same constitutional guaranty apply to the airing of personal disputes and private matters? It seems to the Commission that it does not. Listeners have no protection unless it is given to them by this commission, for they are powerless to prevent the ether waves carrying the unwelcome messages from entering the walls of their homes. Their only alternative, which is not to tune in on the station, is not satisfactory . . . The commission is unable to see that that guaranty of freedom of speech has anything to do with entertainment programs as such . . . The commission believes it is entitled to consider the program service rendered by the various applicants, to compare them and to favor those which render the best service."[53] Thus, by its own Meiklejohn-ian definition of free speech, the Radio Commission claimed to escape the label of censor. Enforcing its judgment about what pattern of entertainment best served the American public was not, in its opinion, a matter of freedom of speech.

By 1929, former Commissioner Bellows was predicting that Congress would "progressively strengthen the power of the Government to investigate the type of service being rendered by broadcasting stations, and to issue or refuse licenses on the basis of the information it secures."[54] When in 1931 the Radio Commission was asked by the American Newspaper Publishers Association to promulgate an order banning the broadcasting of lotteries, the commission declined on the basis that it lacked censorship power, yet three days later it warned stations in a press release that if a substantial number of complaints were received regarding a particular station's lottery broadcasts, it would set a hearing on that station's license application because it doubted that lotteries were in the public interest.[55] Clearly, the commission saw its custody for the public interest as including program content.

One question at the time was how far the commission should go in enforcing "fairness" in the presentation of public issues. In 1930 Senator C. C. Dill, citing the equal time requirement for candidates, asked if the Radio Commission had considered making similar regulations on discussions of public questions: "It would be within the power of the commission," the Senator said.[56] In the same year, the Radio Commission warned broadcasters to exercise more control over advertising. A commission attorney argued that if the commission had jurisdiction to prevent interference in radio reception, it surely had authority to protect the public from "influences of a more dangerous kind . . . Control for one purpose and not for the other is not in harmony with the avowed intention of Congress to regulate radio communication for the best interests of the many. It thus becomes imperative for the commission to be guided by a station's past program record."[57] Slowly but surely the Radio Commission was extending its sway.

While controversy was brewing with respect to such censorship, even greater public concern was focused on the threat of monopoly in the radio industry. Starting in 1926 networks had been established. Of the eighty-nine radio channels available for broadcasting, forty were "cleared" channels, that is, reserved for only one station at a time. Of these, practically all were held by network affiliates.[58] Independent stations were suffering. University stations were dissat-

isfied with the poor wavelengths and low power they had been assigned, and many educational stations were failing, at least partly
owing to the expense of trying to defend their licenses from the
commercial stations, which continued to petition the Radio Commission for more air time.[59] For example, the New York socialist
station WEVD was given low power and relegated to a poor position
on the dial, where it had to share time with eleven other stations.
WCFL, the Chicago labor station, likewise had low power, so that its
reception was interfered with by two Westinghouse stations. A
promise by the Radio Commission in 1928 to grant its application for
a 50,000-watt transmitter remained unfulfilled in 1931.[60]

The courts had their first opportunity to pass on questions of censorship in radio around 1930. In two cases the District of Columbia
Court of Appeals upheld the Radio Commission's refusal to grant or
modify a license and suggested that the commission could consider
the program content of stations in making such decisions.[61] The right
of the commission to consider program content in deciding to issue a
license was posed more sharply in a case in which the commission
had denied a license to Dr. John R. Brinkley, owner of station KFKB
in Milford, Kansas, who gave medical advice over the air and solicited patients for goat gland operations. The station, the commission
held, was not operating in the public interest. In 1931, the same year
that Near v. Minnesota established that prior restraint of the press is
unconstitutional, the circuit court upheld the commission's denial
of a license to Brinkley's station and turned aside his argument that
such a denial was censorship: "This contention is without merit.
There has been no attempt on the part of the Commission to subject
any part of appellant's broadcasting matter to scrutiny prior to its
release. In considering the question whether the public interest, convenience or necessity will be served by a renewal of appellant's license, the Commission has merely exercised its undoubted right to
take note of appellant's past conduct, which is not censorship."[62]

The *Yale Law Journal* pointed out the contradiction inherent in this
case: "Although the Commission may have no power to scrutinize
and reject programs prior to their release, the power to revoke or refuse the renewal of a license is in many cases so effective a means of
censorship as to make unconvincing any legalistic distinction between 'previous restraint' and a refusal to renew a license because
of the character of past programs."[63] The most thorough legal treatment castigated the decision: "This is not something resembling
censorship, it is censorship in fact, the very essence of it." It also crit-

icized the commission's revocation of a Portland, Oregon, station's license because it had not maintained "a standard of refinement befitting our day and generation."[64] A speaker on it named Duncan had used profanities and obscenities.

The landmark case on censorship was Trinity Methodist Church, South, v. Federal Radio Commission in 1932.[65] "Fighting Bob" Shuler, a minister and owner of station KGEF in Los Angeles, had used the station to broadcast such views as economic conservatism, anti-Catholicism, and criticism of local corruption. After the Radio Commission refused to renew his license, Shuler appealed, alleging denial of free speech. The federal appeals court, though acknowledging that the First Amendment prohibited prior restraint on publication, found that the commission's action was a proper exercise of regulatory power: "If . . . one in possession of a permit to broadcast in interstate commerce may . . . use these facilities . . . to obstruct the administration of justice, offend the religious susceptibilities of thousands, inspire political distrust and civic discord . . . then this great science, instead of a boon, will become a scourge . . . This is neither censorship nor previous restraint, nor is it a whittling away of the rights guaranteed by the First Amendment . . . Appellant may continue to indulge his strictures upon the characters of men in public office . . . but he may not . . . demand, of right, the continued use of an instrumentality of commerce for such purposes."[66]

The decision was controversial. Charging prior restraint, the American Civil Liberties Union represented Shuler on an appeal to the Supreme Court, which the Court declined to hear. There was a call for changing the law so as to treat radio just as the press. The Radio Commission was called a "class agency, a political agency without any real freedom."[67] Typical of the concern expressed about censorship was the comment: "While the present state of the broadcasting art does not permit the operation of an indefinite number of stations, this seems to be an insufficient justification for holding that Congress may upon specious grounds restrain utterances which are perfectly legitimate when other media of publication are employed. The regulatory power of Congress should be confined to its proper Constitutional ambit: punishment after the fact."[68]

But supporters of the decision, while acknowledging that prior restraint had occurred, viewed it as justified in the "public convenience, interest, or necessity." According to one commentator: "An analogy . . . of the radio and the newspaper fails, and freedom of the press provision offers no assistance . . . There are only a limited number of radio channels, and there may be an infinite number of newspapers . . . This right [to use broadcast channels] should not be given away free to certain individuals to do with as they choose, and to annoy and harass the public in general."[69] None of the commentators contested the public interest standard itself. The limitation on radio channels was taken for granted, and so was the need for some regulation.

The press, having previously acquiesced in or even encouraged the establishment of controls on broadcasting, now cited the partially censored condition of radio as a reason for not extending to broadcasting the same privileges enjoyed by the press. When in 1933 CBS petitioned Congress for admission of its news people to the House and Senate galleries, representatives of newspapers opposed this request on the ground that "official recognition of radio broadcasting as a medium of disseminating news would be an official sanction of the censorship of news."[70]

The depression and the attendant political mobilization achieved by President Franklin Roosevelt resulted in a new high in censorial actions. The principal targets, rather than being the usual socialists, labor unions, and leftists, were now the conservative opponents of the New Deal. In 1933 Federal Radio Commissioner Harold A. Lafount notified all radio stations of "their patriotic, if not bounden and legal duty" to refuse broadcasting facilities to advertisers who were "disposed to defy, ignore or modify" the codes established by the National Recovery Administration (NRA), the key agency of the early New Deal. The warning was taken seriously by CBS, which canceled a broadcast by Fred J. Schlink, president of Consumers Research, who planned to criticize the NRA. But when some senators threatened to take this issue before Congress, CBS reversed its decision.[71]

Similar restrictions affected foreign policy critics. In 1933 Walter E. Myers of NBC, manager of station WBZ in Boston, wrote the Massachusetts American Legion that in speeches over WBZ the Legion had ignored "one of the important regulations of this company . . . We are obliged to impose regulatory and prohibitory 'rules of the game.' Particularly in a time of national crisis, we believe that any utterance on the radio that tends to disturb the public confidence in its President is a disservice to the people themselves and is hence inimical to the national welfare." And when Walter L. Reynolds of the American Alliance of Patriotic Societies sought time from CBS in 1933 to oppose recognition of the Soviet Union, Bellows, now vice-president of CBS, replied that "no broadcast would be permitted over CBS that in any way was critical of any policy of the Administration."[72] The moratorium on political criticism of the New Deal did not last long. By 1934 the press had turned against Roosevelt. Debate resumed in Congress and, to a lesser extent, on radio.

One New Deal undertaking was a new communications act. The goal was not to reform broadcast regulation but rather to remove telecommunications regulation from the Interstate Commerce Commission and to bring that activity together with radio regulation under a new communications commission, the FCC. The act had two distinct parts, one on telecommunications common carriers and one on radio, the latter being basically unchanged from the Radio Act of 1927.

Debate over the new bill brought out criticism of two types of government interference with radio: the Radio Commission's "censorship by press release," and the alleged domination of radio by the Roosevelt administration. One congressman complained: "When the Government has the power to issue licenses to operate radios it inherently follows that this Government agency has too great a control over freedom of speech . . . So, after all, the real protection of the people yet rests in the freedom of the press rather than in freedom of speech since the coming of radio."[73] In a similar vein a senator warned: "Freedom of speech over the air is at stake when we give the power to license and to destroy. Uncontrolled power to regulate our avenues of communication, our telegraph and cable systems, is the power to throttle the transmission of intelligence with its consequent effect on the press and the public platform."[74]

The press rallied behind a Senate attack on the bill, because as a communications act rather than just a radio act, it would affect the wire services. Senator Thomas D. Schall charged that the bill would permit administration censorship of press dispatches and cable messages. Both the American Society of Newspaper Editors and the Newspaper Guild opposed the powers granted to the FCC.[75] Now that the press had learned to live with radio as a fact of life, it was alarmed by the spread of regulation and censorship. The day was approaching when the press and radio would see themselves as one entity—the media—whose freedom had to be defended.

Despite these concerns and despite the tenor of court decisions, the Communications Act of 1934 perpetuated the established radio system. The system being already in place with an enormous constituency of avid listeners, it would have taken a crisis to prod Congress to try something new, and there was no crisis. So by 1934 a major American medium had come to be licensed, regulated, and even censored. Radio, and later television, suffered a regime encompassing censorship though the law said there may not be any, and in fundamental contradiction to the nation's traditions. Congress had rejected a government-owned broadcasting monopoly, keeping the system pluralistic, competitive, and private, but also under tight and intrusive regulation. Far from there being "no law" abridging freedom of speech, Congress had made such a law and set up the FCC to implement it.

The Courts' Reaction

Court decisions about free speech in broadcasting have wavered, generally supporting the regulatory system but occasionally pulling it back from excesses. For the first couple of decades the Supreme Court ducked the constitutional dilemma for free speech inherent in a system of regulated broadcasting. In 1940 it sustained the constitutionality of the Communications Act of 1934 without so much as a mention of the First Amendment. "Congress," Justice Frankfurter declared, "in order to protect the national interest involved in the new and far-reaching science of broadcasting, formulated a unified and comprehensive regulatory system for the industry."[76] Congress, in his view, acquired the right to regulate a communications industry by virtue of one of those considerations that were presumed to overbalance the prohibitions of the First Amendment, namely a "fear that in the absence of governmental control the public interest might be subordinated to monopolistic domination in the broadcasting field." Licensing in the public interest, which in most countries of the world existed for print but which in America had been outlawed for that medium a century and a half earlier, was rationalized for broadcasting by the goal of protecting the spectrum from monopolists.

Three years later, in NBC v. United States, the Court was more definitive. The FCC, while pursuing Congress' injunction to encourage localism in broadcasting but also pursuing its antitrust policy, had adopted regulations limiting how radio stations could contract with networks. Stations were forbidden to surrender their control over programing.[77] NBC charged that to limit the way stations did their programing abridged their freedom of speech. The Court did not agree. Justice Frankfurter reiterated that because of inherent scarcity, broadcasting was unlike other communications modes: "Freedom of utterance is abridged to many who wish to use the limited facilities of radio. Unlike other modes of expression, radio inherently is not available to all. That is its unique characteristic, and that is why, unlike other modes of expression, it is subject to government regulation."

Twenty-six years later a landmark case, Red Lion Broadcasting Company v. FCC, carried these earlier decisions to their logical conclusion. It sustained regulation not just on the structure of the industry but even on content. At issue was the fairness doctrine, or more specifically, the right of reply to a personal attack.[78] During the 1964 presidential campaign, on a station owned by the Red Lion Broadcasting Company, a fundamentalist radio preacher, Billy Hargis, had attacked journalist Fred Cook for an article criticizing candidate Barry Goldwater. Cook asked for time to reply. In the background, the Democratic National Committee was using Cook's request as a way to intimidate broadcasting stations from carrying Hargis-type attacks. This effort, though irrelevant to the case, is not irrelevant to understanding the consequences of government regulation of speech. The power of the state had become a weapon manipulated by interested parties to silence their opponents.

The FCC ruled that Cook was entitled to free time for a reply. The Supreme Court upheld the constitutionality of thus requiring broadcast stations, in presenting public issues, to give each side fair coverage, and also of requiring them to give persons subjected to personal attack an opportunity to reply. The Court, despite the First Amendment, justified regulation of broadcasting on the grounds that "broadcast frequencies constituted a scarce resource whose use could be regulated and rationalized only by the Government. Without government control, the medium would be of little use because of the cacophony of competing voices, none of which could be clearly and predictably heard." Since only some of those who wished to broadcast could be given broadcast licenses, it was proper that, "Every licensee who is fortunate in obtaining a license is mandated to operate in the public interest and has assumed the obligation of presenting important public questions fairly and without bias."[79]

Citing precedents concerning movies and sound trucks, both of which the Court had dealt with differently from canvassers or publishers, Justice Byron White argued that "differences in the characteristics of news media justify differences in the First Amendment standards applied to them." And so, "where there are substantially more individuals who want to broadcast than there are frequencies to allocate, it is idle to posit an unabridgeable First Amendment right to broadcast comparable to the right of every individual to speak, write, or publish." Rather, "it is the right of the viewers and listeners, not the right of the broadcasters, which is paramount." In consequence, "there is nothing in the First Amendment which prevents the government from requiring a licensee . . . to conduct himself as . . . fiduciary with obligations to present those views and voices which are representative of his community and which would otherwise, by necessity, be barred from the airwaves."

If the premise of an inherently scarce resource were sound, these conclusions would be entirely plausible. If it were true that there could be only a few stations, then freedom would be better served by requiring that they be open forums rather than privileged voices. Indeed, if technology did so limit the number of outlets, the logic of the situation would suggest adoption of something like a common carrier arrangement.

The logic of treating monopolist media owners as trustees is so persuasive that at mid-century some theorists scrapped the traditional print-derived notions of the meaning of the First Amendment in favor of a new theory of "access." The leading such theorist, Jerome Barron, argued that in an era of mass media gigantism, the exercise of free speech by media owners while access to the media is denied to others is not a democratic right. The access movement was an aspect of the dissidence of the 1960s. Citizen groups were formed to force broadcasters to be more responsive to what was viewed as the people's needs. These groups contested license renewals and demanded the right to reply to ads and to programs with which they disagreed.[80]

In general, such access proposals, though they got a hearing, were unsuccessful before the FCC and the courts. One question was whether the fairness doctrine applied to commercials. Initially the FCC supported the demand that antismoking messages should balance cigarette ads, and the courts approved.[81] Conservationists then asked to reply to gasoline company ads. Opponents of the Viet Nam War asked to run ads to which others might then in turn demand the right of reply. The authorities beat a hasty retreat, foreseeing a rising flood of rejoinders to commercials.

The problem of rejoinders to cigarette ads was solved by Congress. The last thing the cigarette companies wanted was to have their ads debated. If that was going to happen, they would rather run no television ads. So with the industry's acquiescence, Congress prohibited all cigarette advertising on the air.[82] The FCC had earlier ruled that cigarette smoking was a special case because of its officially established health hazard and that in general there was no right to balance advertising claims by replies. When the courts seemed disinclined to accept this notion of the uniqueness of the smoking issue, the FCC simply rescinded its ruling on cigarette ads, which Congress had already made moot. The FCC then ruled that the fairness doctrine did not apply to "commercial" ads, only to "advocacy" ads. Since the networks accept "advocacy" ads on public issues only during election campaigns, there was nothing left in advertising material to which to reply.

The Democratic National Committee challenged this outcome. It took CBS to court seeking to compel it to run an ad criticizing a Republican administration's actions in Viet Nam. The Supreme Court on First Amendment grounds supported the right of CBS to refuse the ad.[83] The decision was that although the licensee was required to maintain fairness in its treatment of public issues, it retained editorial responsibility for when, where, and how to carry that out. No third party had a right to have its material put on the air: "For better or worse, editing is what editors are for and editing is selection and choice of material. That editors—newspaper or broadcast—can and do abuse this power . . . is not reason to deny the discretion Congress provided . . . The authors of the Bill of Rights accepted the reality that these risks were evils for which there was no acceptable remedy other than a spirit of moderation and a sense of responsibility—and civility—on the part of those who exercise the guaranteed freedom of expression."[84] The First Amendment, in short, applies to broadcast journalism, though not fully, for it is limited by the fairness doctrine: "A broadcast licensee has a large measure of journalistic freedom but not as large as that exercised by a newspaper. A licensee must balance what it might prefer to do as a private entrepreneur with what it is required to do as a 'public trustee.' " To allow the FCC to mandate the running of any particular ad would bring government into the exercise of "day to day editorial decisions."

In a concurring decision Justice Douglas dissented from the fairness doctrine in general: "The Fairness Doctrine has no place in our First Amendment regime. It puts the head of the camel inside the tent and enables administration after administration to toy with TV or radio in order to serve its sordid or its benevolent ends." Under "the laissez-faire regime which the First Amendment sanctions," the FCC "has a duty to encourage a multitude of voices but only in a limited way, viz., by preventing monopolistic practices and by promoting technological developments that will open up new channels." Television and radio, Douglas argued, stand in the same protected position under the First Amendment as do newspapers and magazines. The fear that Madison and Jefferson had of government intrusion is perhaps even more relevant to television and radio, he held, than to newspapers and like publications, for that fear was founded on the specter not only of a lawless government but also of a government under the control of a faction inclined to foist its views of the common good on the people.

A decisive rejection of the access interpretation of the First Amendment came when its advocate, Barron, argued a case that attempted to apply it not to broadcasting but to the press. Half a century earlier Florida had enacted a law giving candidates for office the right to have published a reply to any newspaper attack on their character or record. The law had been generally forgotten and never tested in court. In 1972 the *Miami Herald* ran an editorial attacking a union official, Tornillo, who was running for the state legislature, and it declined to publish his reply. When the suit reached the Supreme Court, a unanimous decision held the Florida law unconstitutional.[85]

In other countries, such as France, there is a well-established right of reply.[86] Anyone attacked or misrepresented in a newspaper may reply. People exercise this right surprisingly rarely, but when they do, all newspapers but one print the reply. The Communist *l'Humanité* never prints replies, and in the politics of France, despite the law, no court compels it to comply.

In the United States, however, the courts have mostly continued to define free speech as the freedom of the owners of a medium to say what they wish in it. Except for broadcasting, where the opposite conclusion was reached in Red Lion, the courts have rejected the thesis that gigantism in the institutions of expression makes obsolete the notion that owners are the ones free to use their medium. The First Amendment continues to be read as permitting free, robust, and unrestrained expression by publishers rather than as requiring a fair, publicly organized forum.

Broadcasting is, however, an uncomfortable partial exception.

Through licensing and the fairness doctrine, the government administers the debate. Although the courts have on occasion blown the whistle when government seemed to overstep the bounds of permissible intervention, the temptations on the FCC to use its powers to set wrongs right have been palpable.[87] Mothers' groups, religious groups, patriotic groups, and consumer groups have repeatedly called for action against the violence, sex, consumerism, or aberrant values portrayed on the home screen.

The FCC has on occasion ventured far into issues of content. When it required a non-Communist affidavit from merchant marine radio operators, the Washington, D.C., Court of Appeals sustained that.[88] When it restricted the subjects of discussion permitted on amateur and citizen bands, an appeals court sustained that. "Here is truly a situation," Judge Henry Friendly observed, "in which if everybody could say anything, many could say nothing." And so, it is not unconstitutional to prohibit chitchat on the limited available frequencies.[89] When the FCC prohibited a New Jersey radio station from broadcasting the results of the official state lottery, Congress corrected the FCC by changing the law. The Supreme Court then considered the case moot, though Justice Douglas irately dissented: "It is to me shocking that a radio station or a newspaper can be regulated by a court or by a commission, to the extent of being prevented from publishing any item of 'news' of the day. So to hold would be a prior restraint of a simple and unadulterated form, barred by constitutional principles. Can anyone doubt that the winner of a lottery is prime news by our press standards?"[90] And finally, when the FCC warned a station against playing George Carlin's record of "seven dirty words," the Court sustained that repression of profanity. This decision is "a legal time bomb," because Justice John Paul Stevens' argument that broadcasting differs from print was based not on the usual catechism about spectrum shortage, but on the "uniquely pervasive presence" of broadcasting and its confrontation of citizens "in the privacy of the home," which "outweighs the First Amendment rights of an intruder." This aberrant approach could be used to justify quite radical censorship.[91]

The members of the FCC are torn by their regulatory powers. They accept the First Amendment notion that they should not regulate the content of broadcasts, but they are under enormous pressure to control abuses. A Surgeon General's Committee of experts found that violence on the screen was causing antisocial aggression in children. As serious a group as Action for Children's Television asked that drug ads be prohibited when children were watching. FCC commissioners have pondered ways to protect children from violence, pornography, sugar-coated cereals, and cure-all drugs.

At one time, rather than adopting regulations on children's exposure that would turn the FCC into censors, Chairman Richard Wiley called the networks in for a frank conversation and put it up to them to regulate themselves. The result was an agreement to establish a "family viewing hour" up to 9:00 P.M. from which the more violent programs were banned. Producers of displaced programs sued, claiming injury to their business by unconstitutional government pressure to control program content. A district court agreed that the pressure was indeed an exercise of censorship.[92] The FCC has mostly intruded into station programing by the intimation of license nonrenewal, often merely by a lifted eyebrow. In 1976, for example, the FCC held back on renewal for KIBE until the licensee substituted a talk program for a classical music one. The impulse toward growing control is clearly there, only partly checked by occasional vigilance of the courts.

The shortage of spectrum for broadcasting persuaded Congress, the FCC, and the courts that broadcasting was by nature different from print—that it had inevitable elements of monopoly and scarcity. For the print media, even when they became monopolies, as in one-newspaper cities, the decision as to who should be the monopolist was a nonpolitical, evolutionary result of a market process. In broadcasting, there was no time to let that happen. The requirement to do something about radio interference seemed immediate. Political decision seemed inexorable. So people perceived or misperceived broadcasting as a new type of communication system, necessarily different from the old, in which the holder of the physical facilities was a trustee, licensed to serve the public interest and obliged to provide a responsible forum.

It is hard to reconcile such governmentally imposed requirements with the traditional concept of the freedom of the press. The broadcast model assumes that the government has a positive role to play as licensor and regulator. The optimistic notion that government is to play that role on behalf of citizen freedom rather than against it is not persuasive to those who are skeptical about the power of good will in political processes to guarantee good results.

A Rejected Option: A Common Carrier System

Even if it were true that the airwaves are inherently scarcer and less accessible to the public than is the press, still a puzzle remains. Why did the Congress not deal with the problem in the same way that it had dealt with communications monopolies before, namely by the system of common carriage? Why did it instead muddle through to such a confused and uneasy compromise that constituted regulated publishing, a system in which broadcasters have to pander to the powers above them so as to preserve their licenses, all the while waving the flag of their editorial autonomy and First Amendment rights?

Other options were available to Congress in 1927 and 1934 that would have avoided the censorship inherent in a regulated system of broadcasting. One of these, a common carrier approach, had been proposed by AT&T. Called "toll broadcasting," it was rejected, out of fear of adding to the power of the telephone monopoly.

The entrepreneurs who opened broadcasting stations in the 1920s were small and medium-sized businessmen. They saw AT&T's proposal as a major threat to their fledgling enterprises. Researchers today looking at articles and hearings from the mid-1920s may easily misinterpret the frequent alarms about the threat of broadcasting monopoly as referring to the power of licensed broadcasters and networks, as it does today. "Monopoly" in the radio literature of the 1920s, however, was the code word for AT&T and its proposed system of common carrier transmission. This is what Secretary Hoover was referring to in 1924 when he boasted that under American policies, "radio activities are largely free. We will maintain them free— free of monopoly, free in program, and free in speech."[93] In 1926 Congressman Ervin Davis, citing Hoover, added: "We cannot allow any single person or group to place themselves in a position where they can censor the material which shall be broadcast to the public." He pointed to the testimony of W. E. Harkness of AT&T that his company often rejected applications for service. "We take," said Harkness, "the same position that is taken by the editor of any publication."[94]

Between 1922, when AT&T defined for itself the role of a broadcasting common carrier uninvolved in message content, and the 1926 Harkness testimony, which adopted the stance of an editor, AT&T's policy had inadvertently changed. AT&T's radio station WEAF had opened its doors for business, but no one came to buy time on the air. It soon became obvious that either the broadcasting station itself must take the initiative in developing programing, or there would be none. Later, when broadcasting became profitable thanks to large audiences and advertising, a common carrier system might have been feasible. Programers and advertisers would then have found it worthwhile to buy air time. But by that time AT&T was out of the business, and the functioning broadcasting stations had no interest in yielding up the most profitable part of the enterprise to program providers, keeping for themselves only a transmission charge.

AT&T left broadcasting in 1926, selling WEAF—the future WNBC—to RCA in return for RCA's using telephone company lines for networking. AT&T was thus largely out of the picture, but the debates about the 1927 Radio Act reflected continuing opposition to telephone companies doing broadcasting. AT&T was the giant of American industry, while broadcasting companies were still small, struggling, and largely local. The broadcasters, Congress, and the political liberals all joined in the hue and cry against monopoly control of broadcasting. As a result, the Radio Act restricted cross-ownership of telephone companies and broadcasting stations.[95] Broadcasters also did not wish to have common carrier obligations imposed upon them. The industry, in an unpublished memorandum to the Senate, urged "that a broadcaster should not be deprived of his right to refuse . . . advertising or to refuse to render a service at his discretion . . . in exactly the same manner as a newspaper."[96] They were joined by liberals and Congress in not wanting AT&T as the radio carrier. Thus the common carrier option was rejected as implying monopoly; that option came forth tainted from birth by the giant organization which proposed it.

Having separate entrepreneurs doing transmission and programing was normal in a common carrier scheme. Such separation of carrier and content would also have been possible without applying all of common carrier law. Various legal arrangements were available under which one set of organizations could have operated transmitters, leasing or allocating time to other organizations. Such arrangements would have allowed an indefinite number of programers to share a finite number of transmitters.

Today in some countries, such as France and Holland, the broadcasting transmitters are run by one organization and the programing is run by others. This keeps the transmitters loaded while sharing the scarce air time among a variety of program producer groups. When the demand for access by such groups exceeds the time and frequencies available on the transmitter, then either selective licensing, or some rationing scheme, or a price mechanism must be used. If Congress had chosen to consider a common carrier scheme before 1927, it would have had to deliberate about these important details. However, no such schemes for sharing transmitter facilities and frequencies were considered.

Any such schemes separating carrier from content might have proved unrealistic at that early stage of the broadcast market, just as AT&T's common carrier plan did. It is questionable whether investors would then have been attracted to the programing part of the business without the monopoly advantages of possessing a frequency. But these issues were never faced; the only version of common carriage that was discussed was AT&T's monopoly scheme, and this was politically unacceptable.

An Option Not Thought Of: A Market in Spectrum

Another option that was totally overlooked in the early radio debates was for spectrum to be allocated, like paper, ink, and printing presses, by market mechanisms rather than by licensing. A common assumption of either a licensing or common carrier scheme is that the resource being allocated is either monopolized or at least very scarce. The assumption of distribution by a free market is, on the contrary, that scarcity is only moderate and thus manageable by the device of private property and its sale or lease. The policy makers in the 1920s and 1930s, wrongly it now appears, did not believe spectrum was abundant enough to be handled in that way.

Property in something as intangible as airwaves was also a puzzling idea. An understanding of how it could work came about only with articles by Leo Herzel and R. H. Coase in the 1950s, well after the die of radio policy had been cast.[97] In retrospect, designing the broadcasting system to be as much like the existing publishing system as possible seems an obvious alternative to have been considered, but in the early days of broadcasting, the idea of buying and selling rights to use the airwaves did not enter the decision makers' minds. This was evidenced as late as 1958 when a congressman questioned Frank Stanton, president of CBS and the unchallenged intellectual of the broadcasting industry, about the auctioning off of broadcast frequencies. Stanton begged off answering, saying that he had never thought of it before.[98]

Herzel's and Coase's novel proposal was that instead of government giving away frequencies free to recipients chosen by a political process, spectrum assignments should be leased, sold, or auctioned for a market price. Having the government charge for public lands, water rights, or postal service was familiar. Frequencies, however, seemed different. Partly it was a matter of imagery. Nothing physical was transferred to the fortunate radio licensees, only permission to tune their transmitters to a particular frequency. It seemed inappropriate that government should charge for that.

In fact, however, there is a market in spectrum. It is a market in tangible things because what is bought and sold is broadcasting stations. The government initially gives away licenses free; these are then sold in a second-hand market. What is excluded from market allocation is only the initial grant of a frequency by the government to its first "owner." Once licensed, radio and television stations are bought and sold, creating a spectrum market, but a poor one. Theoretically, the FCC has some say when the persons to whom it has once given away a frequency free, on the grounds that they are the best available licensees, sell their license to others, but in general, the approval of such sales is routine. Licenses thus end up belonging to people who were never reviewed but have the money to buy them. So a market for spectrum does exist in resale, even though the initial grant of a frequency by the government is a political decision outside the market system.

Three main advantages arise from having market distribution of spectrum from the start, as opposed to today's system of making political assignments initially that lead into a second-hand market. The first has to do with equity. The present system yields windfall profits to a few individuals, whereas a spectrum market would recapture the windfalls for society. The second advantage is economic. The present system makes for inefficient use of spectrum and thus causes its scarcity, whereas a market system achieves equilibrium by both reducing demand for and increasing supply of usable bandwidth. The third advantage is political. The present system involves the state in licensing preferred broadcasters and censoring from the air those whose values it does not share, whereas markets allocate resources by a game that, like any game, is not always fair but is at least insulated from government. Indeed, a principal advantage claimed for the present regulated licensing system over a market is that it enables government to implement its goals and, specifically, to subsidize the kind of broadcasting it favors.

Under existing practice, the original licensees make a windfall profit by selling the license to someone else; the government gets no compensation for the spectrum it has allotted. The initial licensees are thus subsidized by the public. In television station sales from 1949 to 1970, the average station had an original cost of $994,000 and a sale price of $3,201,000. A 1973 study of economic rents earned by owners of VHF television stations after paying the cost of capital shows a rate of pure profit of 42 to 52 percent.[99] In the past decade the price of stations has zoomed. By now a big-city VHF television station sells for many millions of dollars. A Boston station was recently sold for $220,000,000.

The bonus to the initial licensees does not come entirely free. A queue of pleaders using every political trick and device of influence can be counted on to try to win the original license from the government. Contenders spend money for lawyers and for community backing. The licenses are thus not costless, but the government gets no part of what the bidders spend on public relations. If the market mechanism created for broadcasting had been pushed one level further back and the government had offered spectrum rights for lease or sale at a price reflecting market value, any windfall would have gone to the public, not to politically favored individuals.

The reluctance to charge broadcasters for frequencies in the 1920s reflected solicitude for an infant industry. Congress wanted to promote not burden the new art. Television today may be a gold mine, but broadcasting then was a struggling business uncertain of its future. Listeners were a minority. Advertisers were unorganized and unsure of the payoff; indeed it was still undetermined whether advertising would be the main source of revenue. Hoover and Sarnoff both expressed reservations about the desirability of an advertiser-supported system. To charge for licenses under those circumstances would have been a dubious policy.

From about 1925, spectrum charges would have begun to be rational. Broadcasting prospects were beginning to look clearly profitable, and the airwaves were beginning to be congested. The time had come for a system that encouraged efficiency instead of one that subsidized the squandering of spectrum. Had spectrum been treated as a resource for which licensees made payment, they would have had an incentive to invest in conserving it, to share it, and to use it in a variety of simultaneous money-earning ways. But with a free resource, there was little incentive for broadcasters who already had licenses to squeeze more competitors onto the air by themselves investing in more sophisticated and expensive methods of broadcasting.

The scheme of granting free licenses for use of a frequency band, though defended on the supposition that scarce channels had to be husbanded for the best social use, was in fact what created a scarcity. Such licensing was the cause not the consequence of scarcity. The scheme minimized the motive to resell shared use of a frequency; it gave the licensee every incentive to act as a monopolist and speculator. In the long run a system of spectrum charges would have created incentives to supply the public with broadcasting in ways that would have bypassed or reduced the costs of spectrum to the broadcaster, such as cable television or marketing of more efficient receivers to allow for narrower station spacing.

Spectrum charges would not only have resulted in more efficient use of spectrum, and hence more supply, but would also have raised the costs of broadcasting and thus reduced demand. Implicit in discussions of broadcasting at that time, however, was the postulate that any acceptable technical solution had to be a cheap one for the listening public. A solution that bought higher fidelity transmission, more selective reception, or more frequencies, but at the high cost of new types of receivers, was not perceived as a viable solution. In the United States the rapid expansion of broadcasting, which resulted in there being more radios in this country than in all the rest of the world together, was a matter of pride. The technical approach chosen of allocating the relatively narrow band of the most desirable frequencies to broadcasting kept costs low and thus facilitated this explosion. Use of the new, shorter wave bands would not only have obsoleted existing radio sets but also have required more expensive multiband sets and transmitters at then experimental frequencies. The market notion that, when a resource is scarce, one provides more of it in higher priced ways and thereby restricts the demand did not fit in with a populist notion of cheap broadcasting.

Clearly it was policy, not physics, that led to the scarcity of frequencies. Those who believed otherwise fell into a simple error in economics. They failed to recognize that in the market for radio service, as in any other, when prices are fixed below the equilibrium level, shortages occur, and some rationing scheme must emerge to allocate the scarce resource in nonmarket ways. That the creation of low-cost service was what forced the rationing of broadcasting frequencies was rarely recognized when radio policy was made. Rather, the shortage of frequencies was viewed as an act of nature.

The shortage was also seen as peculiar to radio technology, differentiating it from previous means of communication. In short, missing in 1927 was any realization that radio spectrum was a priceable resource like any other, that its scarcity was a function not of its nature but of its low price, that market mechanisms could create an equilibrium between supply and demand for this resource as for any other, that technology could provide added frequencies in reasonable periods of time but at a price, or that market forces, if allowed to operate, would affect the number of frequencies supplied. There was, therefore, no consideration of using price mechanisms as an alternative to licensing.

The notion that nature itself inexorably required the selective licensing of broadcasters has persisted to the present. It is the core of the 1969 Red Lion decision. Although the Supreme Court noted such technological advances as microwave transmission, it concluded that "scarcity is not entirely a thing of the past."[100] Thus the Court in 1969, like Congress in 1927, saw scarcity as a continuing objective fact, not as an economic disequilibrium arising from policy choices.

By the time of Red Lion it was technically possible to provide as many channels on cable television as consumers would pay for. With cable, the limitations on spectrum are gone. But channels delivered to the public by cable cost more than do a few delivered over the air. The policy of keeping costs down discouraged use of such newer, more expensive technologies and thus increased consumer use of the existing limited broadcasting channels.

A simplistic inference might be that use of more expensive methods of multichannel delivery favors the rich, but it is not necessarily so. While the rich will always be able to acquire more of any resource than can the poor, a well-designed free market may reduce—not increase—that advantage. In a free broadcasting market, relatively impoverished interests would be allowed to buy modest slices of time; that is prohibited today. Now, when the FCC gives a license to a broadcaster, that broadcaster alone is responsible for the conduct of all programing on the channel. License renewal depends on the broadcaster's service, fairness, and balance. The licensee dares not sell hours to others to use as they see fit. But in a market situation, free of the overhanging threat of political judgments on the merit of broadcasts, the leasing of time would be a perfectly natural transaction for licensees to engage in. Interest groups that could not afford to pay for a full station of their own could afford a portion of a station's time.

The result would be much like that which exists in publishing. Entry into print publishing depends upon having skill and capital. To buy a major newspaper or magazine takes a lot of capital, but to buy space in an established publication costs much less. In broadcasting, this privilege of buying small pieces of air time is given only to product marketers and to candidates during campaigns. The FCC has recognized that licensees should not be held responsible for balance, fairness, and service to the public interest during the brief product commercials spotted throughout a station's programs. But anything beyond that the licensees may not allow to fall outside their control. Congress recognized the need of its own members for forced access to the licensee's air time when in 1976 it amended the Communications Act to require stations during campaigns to sell reasonable amounts of time at the lowest rate to candidates for federal office. No one else has this privilege.[101] So spectrum allocation *ab initio* by a market could easily be a less commercial and, for small broadcasters, a less costly, more egalitarian, and certainly more libertarian solution than the present system.

Free licenses for stations that broadcast what the FCC thinks the public wants to see or hear may be a subsidy to those poor who want no more than that, but a burden on those poor who have something they wish to communicate. If one's goal is to maximize the opportunities for all sorts of people and groups to communicate freely, one would design a very different spectrum market from the second-hand one that exists today. The broad idea of using a market may be simple and persuasive, but markets are not all alike. There are many kinds of markets. The slogan "Leave it to the market" has become a cliché of those who have a naive belief that one thereby avoids the need for political decisions. On the contrary, a market is not something that happens by itself. It is something crafted by laws; without them it cannot exist.

Markets are not all alike. The stock market, the market for land, the market for paintings, the market for groceries, the market for education, the market for magazines—each works in a different way. The consumption of most commodities, such as groceries, has very little effect on third parties, while the use of other commodities, such as real estate or radio frequencies, is likely to cause considerable interference and cost for neighbors. This kind of externality may be taken into account in the design of a market. Some markets show decreasing costs with the scale of production and thus lead to natural monopolies, while most show increasing costs as the scale of production rises. Some markets have abundant information, while in others key facts are missing unless disclosure is obligatory. The Securities and Exchange Commission requires a prospectus with issuance of a security; a land sale requires passage of a registered deed; drug packages must include information on dosage and side effects; packaged foods have to list ingredients. In other situations no disclosure is required, but the government itself publishes information without which a market would not work well. The government prints its own laws and regulations so parties to private deals can know their rights and obligations. It publishes tables of tides and maps of navigation buoys and inland waterways on the theory that the social costs of accidents far exceed what navigators would be willing to pay for the information. The optimal setup differs for different kinds of markets.

It makes no more sense to say that use of "the market" will solve the spectrum allocation problem than to say that buildings will solve the housing problem without reference to their design. To devise market institutions that work effectively for spectrum is not simple. So when radio was new in the 1920s, workable ideas for a market in frequencies did not spring easily to mind. Since then, a large range of ideas has emerged for various possible spectrum markets.[102] In one market scheme, spectrum might be sold; in another, it might be leased and thus eventually returned. In one market scheme interference with other users might be prevented by selling only a right to transmit at less than a certain power and in a particular direction; in another scheme without such restrictions the users of a frequency might be made liable for payments to any other owner of a frequency with whom they interfere. One scheme might be for a market regulated by a utilities or communication commission; another for an unregulated one. One market might compel competition by applying antitrust laws and cross-ownership restrictions; another might not. One market might sell or lease frequencies only to firms that operate as common carriers; others might make frequencies available to any company that wants to use them regardless of how.[103]

In one market scheme the whole spectrum might be made available in whatever sized chunks the buyers wish and for whatever use they wish to make of it. An advantage of this scheme is that frequencies being used for low-priority purposes would be bought up and put to higher-priority purposes. If a frequency being used for television was needed more urgently for mobile communication, it would be bought and its use changed from broadcasting. However, this scheme would impose high costs on third parties by reducing the opportunity for standardization of the receiving and transmitting equipment that the public owns.

As a result, the government might well continue to designate large blocks of spectrum for major services, such as television and radio broadcasting or mobile, satellite, aviation, maritime, and point-to-point communication, in much the same way that a city is zoned into commercial, residential, and industrial areas. The justification for so doing would be the same as the justification for zoning: users impose social costs on their neighbors by what they do with their property. A radio set can receive over a limited frequency range; the listener turns the dial to move from station to station. It would be unsatisfactory if the stations serving that listener were intermixed with aviation calls, for example, and scattered all over the spectrum. A scheme that sets aside a block of frequencies, each of uniform bandwidth, solely for broadcasting allows the public to tune in with cheap receivers built to receive those particular frequencies. So a scheme that places specified frequencies on a market only for specified purposes and with specified characteristics lowers the social costs in congested airwaves.

Assignment patterns are made in different ways for different services. For example, a citizen band or land-mobile frequency is shared among many users who need to wait their turn; a broadcast frequency is not. Such uses need to be separated from each other. In a plausible market system for broadcasting, therefore, the government might start with an initial block of frequencies for broadcasting and auction them off. According to economic theory, the size of the block should in the long run be allowed to reflect bids arising from different services. But in the short run, one starts with existing blocks of allocations. Requiring television broadcasters to bid against each other for access to the twelve VHF channels may make sense; it is futile to expect them to bid for a channel that existing receivers cannot receive.

To put the matter in technological perspective, using different modes of multiplexing of users onto the same ranges of spectrum at the same time can cause interference. It may therefore be necessary to set standards for the mode of use within any particular range of frequencies. Market mechanisms can then determine which specific individuals gain the right to make these uses. It may or may not be desirable to assign bands for usage-defined services, but it is clearly desirable to set the technical standards within bands. There is a tendency for any one user community to adhere to common standards. CBers want to be able to talk to each other; ships at sea want to do likewise; and radio listeners would be frustrated if each broadcasting station used a different technology. Thus, restricting bands to compatible technical uses may cause a de facto segregation of services. Whether further compartmentalization serves social purposes is an open question. It may be that one market mechanism should be used to determine ownership and programing within a service, and a different, more sluggish market mechanism should be used to shift hunks of spectrum gradually between services. In any case, as long as rights in spectrum use are allocated by giving different frequencies to different users, market mechanisms can be used to pass those frequencies around.

One treatment of both spectrum and satellite orbit locations has been suggested by Charles L. Jackson to satisfy the need of advanced countries like the United States for many orbital slots soon while also meeting the feeling of developing countries that space belongs to all mankind and should not be pre-empted by a few wealthy nations. In this scheme, each country would be assigned orbital slots and frequencies, but unlike the present scheme, countries would have the right to make long-term leases of their slots and frequencies to countries that needed them now.[104]

Given the difficulty of designing a market system appropriate to the specifics of radio broadcasting, it is entirely conceivable that even had market allocation been considered in the 1920s, it would not have been the system selected by Congress. Such a system, however, was not even considered. The option chosen was a highly inefficient, illiberal one, quite contrary to the nation's traditions. The fact that it was chosen can be understood only in the light of the poverty of alternatives contemplated.

Although congressional interest in spectrum markets is still modest, the idea is beginning to stir. Since 1978 several bills to require payment for spectrum rights have been discussed. None seriously considered so far would establish a true market. Some call for user fees, usually administratively set and well below the market price. Indeed, the commercial broadcasters are beginning to move from opposition to support of such a pseudo-market in which they would pay a small license fee and in return would be relieved of regulation, just as if they were owners of property bought at fair market value. Other bills do call for auctions and lotteries rather than FCC selection among competing applicants, but with severe limitations.[105] Deregulation of radio, in particular, whose spectrum needs are less than television's, is being actively pursued by both the FCC and Congress.[106] Ironically, now that Congress, the FCC, and the industry are gingerly edging toward payments for frequency assignments, some of the conditions that have been premises for some such scheme are changing.

If a market scheme had been adopted in 1927, it would now require redesigning in order to meet the needs of the communications technologies of the 1980s. The dimensions of a scheme for property in spectrum that have been considered up to now include its range of frequencies, the geographic area within which the owners have control of those frequencies, the time periods in which they have the right to use them, and the power at which they are allowed to transmit. These dimensions are appropriate to rights under the conventional frequency assignment scheme created under the International Telecommunication Union and until now used for most radio applications. Users are assigned a frequency to use with time, direction, and power limits. These dimensions may not, however, be the appropriate descriptors of spectrum rights in the years to come.

In the long run, new multiplexing approaches in which signals are separated from each other by means other than frequency bands may make obsolete the notion of defining rights to use the spectrum primarily by frequency assignments. If so, this would call for a different kind of market design. In 1927, frequency division multiplexing was the only game in town. Today, however, there are many other ways of sharing transmission channels among multiple users, including time division multiplexing, in which each user is on the same carrier frequency, but an accurate timing crystal allows each user to transmit for a different fraction of a second. On long-distance phone lines on a T1 trunk, each conversation is sampled eight thousand times a second, and twenty-four speakers share the same carrier frequencies. Spread-spectrum transmission is another way of sharing a common range of frequencies among many users.

The design of a market for these newer multiplexing techniques would differ significantly from a market for frequency division multiplexing, in which each user gets full ownership of some range of spectrum turf and a frequency is held exclusively regardless of whether the owner is transmitting at the moment or not. In a market for traffic that is multiplexed on shared frequencies, people would pay by usage. "Volume-sensitive rates" is the jargon term. Users would pay for the traffic load they put onto the system, whether the fee is for bits transmitted, time connected, or whatever. Some utility, presumably a common carrier, would make the shared channels available and be paid by each user. This is how phone trunks operate. This is how future cable-type systems may operate, whether using cables, optical fibers, or microwaves. Broadcasting has not operated this way at all till now, but it could. The assumption that each broadcaster must have an exclusive piece of spectrum turf may become obsolete. Alternatives that, if adopted sixty years ago, would have given a more efficient, luxuriant, and above all freer broadcasting system will not be the optimal schemes for the next sixty years, in which over-the-air broadcasting may become merely ancillary to cable, videodisks, and broadband switched networks. Whatever "solutions" are considered today should be for these newer technologies into which over-the-air broadcasting will fit as just one low-cost means of delivery.

The government's motivation to recapture some of the market value of the spectrum that it allots would be less for such a common carrier operation. There would be no windfall profit to frequency owners, who now reap their profit by making only partial and inefficient use of their frequency while holding it speculatively for later resale. With a common carrier the social benefit of giving away spectrum use free is likely to be passed on to the public. If the government wished, however, it could, by collecting a royalty, recapture some of the value of spectrum that common carriers were making available under usage-sensitive sale schemes.

Even if such radically new schemes never come into play and broadcasting continues to use conventional frequency division schemes, over-the-air broadcasting will no longer be the only way to send video programs to the public. Over-the-air transmission is still, however, by far the cheapest means. To put out programs this way rather than over cable systems or by disks or tapes is thus a privilege that is made available to only a few of the competitors. The question is whether they should pay the value of this privilege. Classical economics tends to say yes, the grounds being rational resource allocation. Furthermore, markets not only optimize resource use, but also remove the hand of government as a distributor or denier of privilege. From that perspective, too, the mechanism of price is a better way than administrative selection to reduce the assignments of spectrum to what is available; it avoids having the weeding out done by political favor.

The First Amendment and the Choice Made

To the detriment of freedom, however, neither the option of common carriage nor that of the market was chosen for radio and television policy. American radio and television has settled into a regime regulated by the values and judgments of public authorities. Apocalyptic prophecy might have projected a trend toward a full dictatorship of the airwaves, but life is more complicated than that. The political selection of broadcasters, when carried out in a pluralistic society with a free printed press and strong traditions of private enterprise and freedom, and governed by a radio law with explicit injunctions against censorship, has produced an uneasy compromise: a system in which political officials meddle, though with reluctance, in the activities of individual stations, and do in fact decide what type of broadcasting the American public wants and shall receive.

By adopting a licensed, advertiser-supported, limited-channel broadcasting system, America has penalized itself for half a century. It has undermined its tradition of free communication, and it has limited broadcasting to mass provision of the few most popular formats of entertainment. The "vast wasteland" of television programs is effectively closed to any video production that does not fit the few molds that attract the largest audiences.

The coming of cable, along with videotapes and videodisks, may ameliorate the problem of narrow uniformity. There is already a nascent market for novel video services. In the past, when Americans got only that video for which advertisers would pay, they got a large and highly polished supply of a limited standard product. American advertisers spent $8.85 billion in 1978 in support of television and $3 billion in support of radio. It cost them far more, $15.3 billion, to support the much more interesting and varied newspapers and magazines. In addition, those print publications got another $6.1 billion from the customers who bought them, bringing their revenues to about double that of broadcasting. Diversity and quality do not come free.

Now for the first time, with pay television, cassettes, and disks, those consumers who want a different product from what Madison Avenue provides and the FCC willingly licenses can exercise some choice by paying for it. Educational courses, ideological propaganda, pornography, religion, high culture, and whatever else substantial groups of people desire to watch are increasingly available through the new media. In the natural evolution of the market, video may begin to offer the equivalent of the little magazines, the special magazines, the cranky magazines, and the serious magazines that the reading public supports.

With regard to the freedom that public policy will allow to these media, however, there is less reason to be sanguine. Institutions change more slowly than markets. The mere growth of new media will not reverse the precedents that were set and frozen into law in the early years of radio communication. Indeed, there is a strong tendency to carry over to the new media, which do not suffer from the special constraints of spectrum shortage, unnecessary and ill-considered precedents of regulation that were set solely on the illusory basis of a supposedly exceptional scarcity for broadcasting. This is a reason for concern.

The time has come to bury the old cliché that spectrum is a scarce resource. It is an abundant resource, but a squandered and misused one. Like any resource, it is limited, but like such other communications resources as paper, trees, printing presses, wires, or television sets, it is plentiful.

As economists use the word, almost all resources are "scarce." That is to say, utilization of a resource withdraws it from alternative uses. Information is an exceptional resource for which this is not true. Giving information to one person does not reduce the amount of it available for another. Spectrum is not like that, but is like water, paper, or petroleum in that use of it by one person limits what is available to others. In that technical sense it is scarce. But in the layman's sense, in which water is abundant but diamonds are scarce, spectrum is abundant. If it were sold at its true price, spectrum would turn out to be a bargain; there is much of it to be had. It is renewable: when a user has used it for an hour or a nanosecond, it is there again afterwards for the next person to use. And what is more, technological progress over the past half-century has multiplied its availability.

Spectrum shortage is thus no longer a technical problem but only a man-made one. The Court in Red Lion was wrong. It looked forward to the day when technology would solve the problem. It did not understand that technology had done so already. What is lacking is a legal and economic structure to create incentives to use extant technologies in ways that would provide broadcasting in abundance.

Like most resources, spectrum is subject to increasing costs to scale, that is, the cost of marginal units gets higher as more of them are supplied. The technologies to provide many broadcast channels cost more than the technologies that use the spectrum lavishly and therefore provide only a few usable channels. The marginal cost of each additional set of channels is likely to be successively higher. At some point the public will vote with its dollars that more channels are not worthwhile. But the failure to create institutions that allow the public to acquire channels up to the point where they choose to stop paying for them is a policy failure, not a technical one.

Methods for Multiplying Channels

There are numerous ways to expand the channels that are available for electronic communication. The methods for doing so include storing the message in electronic memory to be delivered when convenient, tighter channel spacing and localization, improvement of receivers, allocating new frequencies, compression, multiplexing, and use of enclosed carriers.

Unlimited capacity can be had in the form of physically portable electronic memories, such as videodisks and videotapes. Material that would otherwise be broadcast on the airwaves can be delivered to the viewer's premises, stored there, and used when wanted. The delivery is slow, so for some uses these methods are not equivalent to on-line channels. While disks and tapes are the functional equivalents of books, which store large amounts of continuously valuable information, on-line systems using coaxial cables or optical fibers are the functional equivalents of newspapers or magazines, which deliver current intelligence to the public.

Channels for immediate and simultaneous delivery of broadcasts can also be multiplied. Stations could be spaced closer together. In 1980 the FCC considered increasing the number of radio stations by reducing the bandwidth of each from 10 kHz to 9.[1] The broadcasters objected; they did not want more competitors. The FCC dropped the idea.

Stations could be operated at lower power, reducing the interference between stations. If pushed far, this measure would require set owners to purchase better receiving sets. The FCC recently considered licensing a few hundred new low-powered television stations that would be receivable by existing sets. There were thousands of applicants for the licenses, and again opposition arose from the broadcasting industry. Nonetheless the FCC is moving forward with the proposal.

More stations could be had by allocating new spectral bands to broadcasting use. For television, which requires much bandwidth, this option is limited, but for radio it is easy. There are now about nine thousand radio broadcasting stations in the United States, with over sixty stations each in such metropolitan areas as Los Angeles and Chicago. There could be a thousand radio stations in every metropolitan area.[2] The reallocation of two UHF television channels to radio could allow this to happen. The 6 MHz of a television channel would divide into 667 9-kHz voice channels. At UHF frequencies radio waves do not follow the curvature of the earth but stop at the horizon, so these same frequencies can be reused in all but closely adjacent metropolitan areas. To receive these higher frequencies, listeners would have to purchase new sets. And to make it practical to choose among a thousand stations, the sets would need digital tuning; the listener would find the desired station by keying in its number.

Additional television channels could be provided by using microwave frequencies and higher. At high frequencies much more bandwidth is available than at low, because the waves cycle more often. If each wave is thought of as a way of sending one bit of information, then more bits can be sent in the same time when waves go like this ∿∿∿ than like this ⌒⌄⌒⌄. Pay-television broadcasting at microwave frequencies is starting to be authorized by the FCC using what is called "multipoint distribution service" (MDS). While microwaves can be sent in narrowly directional beams, rooftop to rooftop, such distribution of broadcast signals to the entire population is not likely to be cost-effective compared to omnidirectional transmission such as MDS.

Broadcasting directly from satellites to homes permits additional television channels to broadcast, first in the 12–14 GHz band, which is now allocated for that purpose, and perhaps later in still higher bands. There are active plans for such broadcasting in France, Luxembourg, Germany, the United Kingdom, Scandinavia, and elsewhere. Since 1981 the FCC has authorized the first eight licenses for direct satellite broadcasting in the United States.

More messages can be sent in the same bandwidth by using compression techniques. To convey a page containing 250 words by typing in each character at a keyboard and transmitting each as a code of ones and zeros takes about 12,500 bits. To send a facsimile of the same page may take a million bits, that is, a million reports while the scanner passes across the page as to whether a spot scanned is white or black. Because the overwhelming majority of the spots are white, the facsimile machine can be programed to assume a continuing string of bits of the same color until told there has been a change;

that way much less information has to be transmitted. This coding scheme, as well as more sophisticated ones that not only look left and right but also up and down, allow a facsimile page to be transmitted in about 100,000 bits. Although this is still an inefficient method for communicating 250 words of straight text, the facsimile also conveys pictures, different type fonts, and spacing in addition to text.

The compression of a videopicture from its usual 6-MHz bandwidth down to 1.5 MHz allows four television pictures to be sent in the spectrum space now used for one. Since most "pixels" or dots in a picture remain the same between any two frames, picture compression can be achieved by transmitting information only about those pixels that change from frame to frame.[3] If the hero kisses the heroine or the newscaster reads the news, very little in a picture changes in a sixtieth of a second. Use of picture compression is being planned by over-the-air pay television stations to get more channels in their limited frequency quota.

Great progress may be expected in the art of compression. It is most feasible with digital transmissions, in which a processor codes the original at one end and reconstructs it at the other. So compression is likely to be implemented only where the sender and receiver share an incentive to invest in expensive, sophisticated equipment to economize on communication costs. An advertiser-supported broadcasting industry has no such incentive; it is content to squander free spectrum so as to reach large numbers of cheap receiving sets.

All these methods of adding to channel capacity are minor compared to what can be added by taking the signals off the air and sending them in an enclosed medium, such as cable. By this means channels can be increased by orders of magnitude. Cable television systems with over a hundred channels are now being built. The type of cable and amplifier on it that is currently standard can carry fifty-four channels, so a twin cable system can provide one hundred and eight channels. By simply adding cables, one can provide as many channels as consumer demand will pay for. Later on, optical fiber systems may replace ones using coaxial cables, and these will allow for still more and cheaper channels.

Cable

For the present and the immediate future, cable television is the major breakthrough in overcoming spectrum scarcity. The squandering of 6 MHz of over-the-air spectrum for each television channel can be eliminated entirely if policy so ordains. The airwaves can be reserved for those applications for which it is impossible to connect the sender and receiver by a physical wire, such as communicating with moving vehicles or satellites, radio astronomy, and radar. Broadcasting to homes can be provided in unlimited amounts via coaxial cable or optical fibers. These cost more than the airwaves, but in large systems not exorbitantly more.

By the late 1970s millions of families worldwide already received television via wires or cables:[4]

Country	Number of subscribers	Percentage of TV households
Austria	50,000	2.5
Belgium	1,700,000	64.1
Canada	1,326,000	57
Denmark	800,000	50
Finland	50,000	3
France	6–8,000,000	37
Ireland	666,000	23
Netherlands	2,000,000	55
Norway	250,000	22.7
Sweden	1,400,000	46
Switzerland	680,000	36.8
United Kingdom	2,546,000	14
United States	17,400,000	22.4
West Germany	8,000,000	35

By May 1982, cables delivered television to 30 percent of the television households in the United States, for a total of 24 million subscriber families. The figures above mix apples and oranges, however, for the conduit carrying the signal may be a coaxial cable having the capacity to deliver twelve to fifty-four different multiplexed signals, including some not available over the air, or it may be "twisted pairs" of wires with the capacity of delivering only one to four channels. The low-capacity systems are used to avoid having forests of unsightly rooftop antennas or to improve the picture in areas of poor reception without adding signals to what is broadcast. The British, French, and German systems are generally of that kind. The more commodious cable systems, such as those in the United States, Canada, and Belgium, offer a capacity for adding signals not available off the air.

Experience in places such as Canada, where half or more of the households subscribe voluntarily to cable television at prices that fully pay for the system, demonstrates that an entirely cabled nation, with no broadcasting over the air, is not impossible. If virtually all Canadian households passed by the cable became subscribers, instead of only about half of them, the cost per household would be substantially less than that paid today. A few cablecasters are beginning to explore the option of connecting every single home to their cable free of charge, so as to position themselves to make more money on extra services that can be marketed only to connected homes.

An all-cable television broadcasting system with no over-the-air broadcasting is thus not an absurd notion, though it is not necessarily a good one. Removing the most popular mass medium from the cheapest mode of delivery, namely the airwaves, may be a poor use of resources, particularly since reaching remote rural locations by enclosed carriers rather than over the air is expensive. The decision as to what to broadcast over the air and what to send by wire or cable, whether it is made by regulators or by a market, involves trade-offs. There are alternative uses for the spectrum and alternative modes of delivering the message. The optimal solution will vary with circumstances. What is now beyond doubt is that enclosed carriers can be used to provide whatever amount of spectral resource consumers need, and at costs within the range of consideration.

The Government's Thumb on the Cable

In all countries, cable television's advocates have had to combat established broadcasters for the right to create cable systems. Authorities everywhere have made the installation or expansion of cable television either illegal or difficult. In most countries, for example, no cable may be run across a street without government permission. A master antenna within a building or within a block may be legal, but to run a wire across the road infringes the telecommunications monopoly.

In the United States, regulation of cable television is divided between the local franchising authorities and the FCC. City streets are under local jurisdiction; cables are strung along them under franchise. This gives municipalities the leverage to seek either fees or free services from the cable system. Local governments have paid little attention to program content, however, because this has mostly been standard television programing, a matter over which the FCC has charge.

The FCC has gyrated widely in its cable policies. Until 1965 it declined jurisdiction over community antenna television systems (CATV), thus creating a favorable environment for their initial growth. By then, however, the growth of CATV, most notably in San Diego, had begun to pinch local television stations. San Diego, with its poor reception because of deep valleys and hills, is 140 miles from Los Angeles' dozen television stations, which is too far for direct over-the-air reception but a modest distance for importation of signals by microwave. Those conditions favored extensive cable subscription, and as a result, Los Angeles stations on CATV drew audience away from stations in San Diego. To stop this, the FCC ordered a freeze on extending cable to new subscribers. The FCC generalized the freeze by banning cable television systems in the hundred major television markets from importing "distant signals." By the start of the 1970s the FCC decided that it no longer wanted to stop the growth of cable television; it wanted to encourage growth, though in a controlled way. To this end it adopted a new set of rules for cable.

The FCC's 1972 rules may have been pro-CATV, but they were not pro-freedom of CATV. They covered four main areas:

Signal carriage, i.e. the broadcast television signals that cablecasters were or were not allowed to show on their systems, including imports from other cities

What cable systems were required or permitted to offer besides television programs picked off the air, including program origination, access channels, and channels available for lease

Technical standards for equipment and transmission

Division of responsibility between federal, state, and local governments[5]

First Amendment issues fall mainly in the first two areas. In the area of signal carriage, since 1962 there had been rules about "non-duplication." These banned the showing of a program via an imported signal when it was already available on a local station. Such rules were sustained in court in 1968 as not violating free speech.[6] Encouraged by judicial support, the FCC devised elaborate rules about the signals that cable systems may or must carry. Cable television systems were required to carry all local television broadcasts and were allowed to import additional signals, but only in accordance with the rules.[7] These differed for big cities, small cities, and rural areas. For the fifty largest television markets a strictly ranked order of priority governed the stations that might be imported. The first priority was to ensure that all three commercial networks were available. Cable systems were also permitted to provide three independent stations, at least one of which had to be UHF, plus any state-operated educational station, any other noncommercial stations, and any specialty stations with foreign language or religious programing. There were additional permissions for specific programs, such as network programs that the local network station chose not to carry, or late-night programing after the local broadcasting stations had signed off. There was also a requirement to black out home games that had not been released for local broadcasting and certain other programs for which exclusive rights had been sold. The rules barred cable systems from showing motion pictures which were more than three but less than ten years old, except for those picked up off the air. Because the FCC desired that cable systems become a new source of entertainment and information, the rules required large cable systems to originate programing. Franchisees were also to provide "access channels" on which community groups could put their own programing and were always to have enough extra channels available for lease to those who wished to put on their own material.

The new rules were greeted with enthusiasm as a declaration of war on the monotony of over-the-air television. Community groups, poor organizations with causes to promote, noncommercial film makers, and ethnic minorities who felt that they lacked adequate voice on the media saw in local and access cablecasting the prospect of at last getting their hands on video time. The rules not only assured them time but also required the cablecaster to provide studio facilities at modest cost or, for some purposes, free.[8] But soon disillusionment followed.

By 1974, even though the Supreme Court had in the meantime sustained the FCC's right to order local origination, the FCC dropped that requirement in favor of a stronger access requirement.[9] The FCC stopped requiring local origination partly because the audiences were disappointing, but mainly because in the hard times then faced by cablecasters the cost of programing was a burden on the growth of cable. Programs that cablecasters voluntarily originated continued to be subject to the same fairness requirements as applied to broadcasters. On access programs that the cablecasters did not otherwise control, they were responsible for excising lotteries, obscene material, or sponsored material lacking identification of its sponsor.[10]

When in 1969 the FCC applied to CATV's local origination programing the requirement of offering equal time to rival political candidates, the American Newspaper Publishers Association saw a dangerous implication for the future. If in the future newspaper publishers delivered news over cable channels, the equal time regulation would apply to them, they protested. The FCC, squirming to reconcile its content control of CATV with the free tradition of print, replied: "We did not intend to apply these requirements to the distribution of printed newspapers to their subscribers by way of cable ... However ... ordinary cablecasting is covered by the rules. It makes no difference that a newspaper is the originator of the programs any more than it would if a newspaper sponsored a program on the broadcast station ... We have no intention of regulating the print medium when it is distributed in facsimile by cable, but we do hold that the publication of a newspaper by a party does not put it in a different position from other persons when it sponsors or arranges for the presentation of a CATV origination which does not constitute the distribution of its newspaper."[11]

The 1972 cable rules, though now largely dead, illustrate how far, if permitted, regulators will go, with the best of motives in the freest of countries, to control the contents of a medium. These rules have recently fallen before the twin assaults of deregulatory policy within the FCC and overturnings by the courts, but for a while they were enforced on cablecasters. The deregulatory steps taken voluntarily by the FCC were largely in the nature of lifting costly requirements from the shoulders of the industry. Not only was the requirement for local origination dropped, but so was a prohibition against advertisements on locally originated programs.[12] The rules that forbade importation from a distant station of programs available on a local television station were weakened and, in 1980, lifted.[13] The rules designed to prevent a cable system from importing programs from more remote cities and skipping over nearer ones, called leap-frogging, were largely cut out.[14] The requirement for a station to maintain one channel for public access, another for use by local schools, still another for local government, and one more in reserve for lease by others beyond those in current use, was collapsed down to one channel for all those purposes.[15] The scheme that cable enthusiasts had perceived as a grand design for public service cable, with channels made available free to community groups, was cut to a modest set of requirements that would not cost a struggling cablecaster dearly. None of this deregulation can be interpreted as sensitivity to First Amendment issues. The economic concerns of the cablecasters were heard by the FCC, but no rules were expunged on the grounds that the FCC had no business regulating speech.

Does the FCC Have Jurisdiction over Cable?

The courts, however, were not totally supine. Though they gave the FCC a long leash, in bursts of occasional vigilance they puzzled about where the limits of its regulatory authority might lie. Early decisions seemed to give the FCC almost unlimited power over cable systems. Later decisions began to question that authority and over-turned a number of the cable rules. Yet the courts have not seen fit to deny the right of the federal government under the Constitution to control cablecasting. They have puzzled about the First Amendment but then ended up throwing out one rule after another on other than free speech grounds.

A pernicious 1943 *obiter dictum* by Frankfurter about the "comprehensive mandate" of the FCC and its "not niggardly but expansive powers" has come to permeate dicussion of communications regulation.[16] He had earlier interpreted the intent of Congress in creating the FCC to be "to maintain, through appropriate administrative control, a grip on the dynamic aspects of radio transmission" and to formulate "a unified and comprehensive regulatory system for the industry."[17] Out of such general phrases, Frankfurter spun a fanciful picture of a drastic and authoritarian congressional intent in an area where Congress' power is less than for ordinary legislation.

This rhetoric, and other claims of capacious power for the FCC, were based on broad preambular phrases in the legislative history.

The Communications Act of 1934 specified its application to be to "all interstate and foreign communication by wire and radio" and gave the FCC the responsibility to "make available . . . to all the people of the United States a rapid, efficient, nation-wide and world-wide wire and radio communication service." Even if this wording justified Frankfurter's tenuous interpretation, all that would follow from it would be an imperative on the Supreme Court to correct that overbroad encroachment on constitutional freedom. Certainly nothing in the congressional record suggests any deviation from the standard rule that "the public welfare cannot override constitutional privileges."[18]

In any case, these imperial claims for FCC power concerned broadcasting over the air, not cable. But since precedent is the style of Anglo-Saxon law, the courts define a new technology as a special case of a familiar one. Just as telephones were defined as telegraphs, so cable television was defined as television. Yet from a constitutional point of view nothing could be more different than cable television and television. Television, as broadcast over the air, uses "scarce" spectrum which the government rations; cable television uses none.

The confusion of cable television with television began with the San Diego controversy. When the FCC ordered a freeze on cable extensions, the San Diego operators appealed to the courts, but they lost when the Supreme Court ruled in 1968 that the FCC could regulate cable to the extent that it was "reasonably ancillary to the effective performance of the Commission's various responsibilities for the regulation of television broadcasting."[19] The basic argument of Justice Harlan was that the FCC "has been charged with broad responsibilities for the orderly development of an appropriate system of local television broadcasting." Broadcasting and cable television were a "stream of communication" which was "essentially uninterrupted and properly indivisible."[20] The FCC had "reasonably found" that the achievement of its purpose of fostering television was "placed in jeopardy by the unregulated growth of CATV." The First Amendment question as to whether government may regulate one medium of expression just because that medium in some way causes problems for a regulated medium was not confronted. So since that decision the rationale for federal control on cable has been its relation to television.

The relation of cable television to over-the-air broadcasting is in fact complex. In different aspects it is adversarial, synergistic, or cooperative. There are issues of copyright that arise when cable systems relay broadcasts, and there is marketplace competition when cable systems orginate different programs. Broadcasters claimed that cablecasters who picked up broadcasts and redisseminated them to subscribers were violating their copyright. Such a claim for rights in a broadcast over the air could not have been sustained under the common law, which limits copyright to visible texts, but under the copyright statute that Congress had adopted, there was plausibility to the claim. However, in 1974 the Supreme Court ruled in favor of the cablecasters.[21] At issue was the question of whether, using the analogy of broadcasting, a community antenna system was a transmitter of programs or part of the receiver for them. The Court interpreted CATV as simply an extended antenna that helped the subscriber pick up a signal which had been broadcast for anyone to receive without charge. The cablecasters were thus freed of any copyright liability; they were not transmitting.

Two years later Congress changed this situation by revising the copyright law. It established both royalty obligations on cablecasters for retransmission of broadcasts and compulsory licensing by broadcasters. Since 1978, cablecasters who relay broadcasts must periodically file a detailed listing of what broadcasts they have carried and "must pay an aggregate fee calculated on a graduated basis in accordance with annual subscriber revenues . . . Such fees are ultimately disbursed to copyright holders nationwide."[22] Because the fee is low, cablecasters are happy to have the Damocletian sword of copyright removed from above them; but broadcasters are not happy, for they receive very little. They are now campaigning for a change in the law, to eliminate compulsory licensing and compel the cablecasters to negotiate a deal for every copyrighted program they wish to carry. The issue is a difficult one; it is not obvious what a well-considered copyright law should say. Cable spokesmen stress that broadcasts are put out free, for all to receive at will, and that negotiating rights to every half hour of programing on several stations would be an intolerable burden. Broadcasters argue that cable is just another distribution system. It gets its pay programing in the competitive market, and it should get all its programing in this fashion, without governmental intrusion through a compulsory license.

Under the constitutional copyright provision, the solution is in the jurisdiction of the Congress to decide, but whatever the resolution may be, barring congressional authorization, the fact that cable systems relay copyrighted broadcasts is no basis for the exercise of jurisdiction by the FCC. The copyright issue is between the cablecaster and the copyright owner. If replaying a broadcast on cable television violates a copyright, that is a matter for a civil suit in the courts, not for the creation of a system of prior licensing and government regulation. In copyright there is no basis for the FCC to extend its jurisdiction to include cable television. Certainly no court would sustain a claim that the press can be subjected to licensing just because it transcribes radioed material. The fact that cablecasters carry material which originated as broadcasts may be a basis for suits by those whose rights are infringed, but under the narrow mandate that the First Amendment requires, it is no basis for regulation.

The argument for FCC cable jurisdiction has been that the FCC is responsible for the protection of broadcasting. An appeals court concluded in 1968 that the FCC's "effort to preserve local television by regulating CATVs has the same constitutional status under the First Amendment as regulation of the transmission of signals by the originating television stations. It is irrelevant to the Congressional power that the CATV systems do not themselves use the air waves in their distribution systems. The crucial consideration is that they do use radio signals and that they have a unique impact upon, and relationship with, the television broadcast service. Indiscriminate CATV development, feeding upon the broadcast service, is capable of destroying large parts of it. The public interest in preventing such a development is manifest."[23]

But the interest of protecting broadcasting hardly overrides the First Amendment. Surely some vested interests may be injured. The rise of magazines hurt book publishing. The rise of television hurt movies. But no one suggested that those situations required Congress to make exceptions to the First Amendment.[24]

Neither the notion that cable television is broadcasting because it frequently carries the same content forward nor the notion that it is a competitive threat to broadcasting persuasively justifies imposing government regulation over a medium that makes no use of the spectrum. Without compelling and concrete arguments, the advocates of regulation have tended to mumble a certain magic formula. The FCC claimed, and courts in the 1960s tended to agree, that despite the absence of any congressional authorization to regulate cable systems—for cable did not exist at the time of the Communications Act of 1934—and despite the lack of any special feature of cable which makes traditional First Amendment concepts difficult to apply, the FCC had the right to regulate cable television as "reasonably ancillary to broadcasting."[25]

Unease with this position began to be manifest in court decisions in the 1970s. In one key case, United States v. Midwest Video Corporation in 1972, the Court barely upheld the FCC's requirement that large cable systems originate some programing and warned that this requirement "strains the outer limits" of the FCC's statutory authority.[26] In 1977, a court of appeals overturned the cable rules concerning what movies and sports events could be carried on pay cable. No longer were motion pictures three to ten years old banned from cablecasting. Fundamental in this decision was a sharp distinction between cable and over-the-air broadcasting. No longer was cable treated as just an extension of broadcasting. These rules that the court upheld as applied to over-the-air pay television, it overturned as regards cable television.

This distinction was based on a supposed difference between the type of monopoly involved in each medium. Broadcasting, the court believed, as did Congress in 1927, involved a physical scarcity of channels. Neither cable television nor newspapers were said to suffer from such a physical limitation. Monopoly, if it existed in cable television or the press, was merely the result of economic forces. And resting on the precedent of Tornillo, the court argued that "scarcity which is the result solely of economic conditions is apparently insufficient to justify even limited government intrusion into the First Amendment rights of the conventional press . . . and there is nothing . . . to suggest a constitutional distinction between cable television and newspapers on this point." Although the distinction between physical scarcity for broadcasting and economic scarcity for cable was dubious at best, the court seized upon it as a means of differentiating the two media.[27]

In 1978, a second Midwest Video case came to a federal appeals court. In a manner contrary to judicial norms, the court spent pages demonstrating that the requirements of the cable rule to provide access channels for community use violated the First Amendment, but then proceeded to say that "it is unnecessary to rest our decision on constitutional grounds and we decline to do so." The access rules were overturned on the grounds that since no such rule could lawfully be applied to over-the-air broadcasters, the FCC could not justify applying that stringent requirement to cablecasters as ancillary to broadcasting.[28]

In *obiter dicta* about the First Amendment, the court this time argued that cable was not just a community antenna for picking up broadcasts. By 1978 it had become obvious that cables could carry traffic that had nothing to do with what was broadcast through the air. The court distinguished "cablecasting" from "retransmission" of broadcasts and equated cablecasting with print in its status under the First Amendment: "We have seen and heard nothing in this case to indicate a constitutional distinction between cable systems and newspapers in the context of the government's power to compel public access." The court regarded cable operation as "electronic 'publication' " and argued that no one would apply the same sort of rules to newspapers: "Though newspapers 'retransmit' hundreds of government press releases, we assume that no government agency has the fatal-to-freedom power to force a newspaper to add 20 pages to its publication, or to dedicate three pages to first-come-first-served access by the public."[29] The fact that cable systems retransmitted broadcast signals was no reason to extend federal control to their own cablecasting activity.

The decision was contemptuous of the role of access in free speech. Following the tradition of the print media, in which historically access was no great problem, the court saw freedom as essentially the freedom of the media owner. The FCC, the decision said, failed to make any showing that cable systems had been "public forums," that is, common carriers. "If they are not, it would appear that the present access rules cannot withstand constitutional muster." Except in a public forum or common carrier, the media owner's rights were seen as paramount: "The First Amendment rights of cable operators rise from the Constitution; the public's 'right' to 'get on television' stems from the Commission desire to create that 'right.' "[30]

The gyrations of the courts in dealing with cable systems reflect a legal system based upon precedent. Past confusions and errors cannot easily be overturned or disregarded. The courts have reversed various of the more egregious FCC regulations of cable one by one. But given how strongly ensconced in precedent are the notions of broadcasting as regulatable because of the physical shortage of spectrum and of cable as an extension of television, the courts have continued asserting the legitimacy of cable regulation in general, despite the First Amendment. They have evaded the logical conclusion that the FCC has no authority whatever to regulate what may be carried on cable systems and cannot have such authority under the

Constitution. They have so far sustained the FCC's jurisdiction in principle, while overthrowing specific regulations as exceeding the FCC's authority.

The direction of movement in the last decade both in the FCC and in the courts has been toward reduced regulation, the jargon word for which is "deregulation"; but there remains a residue of precedents that severely erode the hard-won principles of a free press. The print tradition is not one of deregulation; it is outlawry of regulation. The FCC for the past decade has prided itself on repealing unnecessary rules. Its philosophy is that of a benevolent ruler, who wields authority lightly and avoids regulation where possible. But every deregulatory decision by the FCC has at the same time reaffirmed its authority: it has merely found certain regulations to be unneeded. Similarly, the courts have in most major recent decisions reduced the scope of cable regulations. But they too have let stand assumptions about the legitimacy of content regulation that derive from problems of early broadcasting.

There is nothing about spectrum technology that today mandates bureaucratic control of what is transmitted by cable. There need be no scarcity of capacity or access. The law that governs print can become the law that governs those who communicate on cable channels too. Since new technologies allow for broadcasting without utilizing the airwaves, broadcasters can enjoy the same freedoms as do publishers. The new electronic technologies, like the old ink ones, can enjoy a system governed by individual choices in the marketplace instead of one governed by political choices made by people the President appoints and Congress confirms. But all of that is true only if those who wish to publish over cable can get onto the cable. The problem of access may become the Achilles heel of what could otherwise be a medium of communication every bit as free as print.

The Future of Broadband Common Carriers

In the coming decades, as cable evolves from a novelty to a major communications utility, a great policy battle will be fought about the organization of the industry, specifically about competition and monopoly within it, and about common carriage versus content control. All the disputes and court cases concerning cable television discussed so far have been about a home entertainment medium. This is what cable has been in the past, but this may not be its future.

As the penetration of cable systems grows and as cable technology is put to innovative uses, coaxial cable networks become no longer just community antennas for entertainment television but rather a broadband delivery system for all sorts of electronic traffic. Cables or other enclosed broadband carriers will continue to bring entertainment to homes, but they will also be carriers for computer data, electronic mail, videotex, information bases, education, security monitoring, teleconferencing, and news services. Their use in maturity may resemble their early use no more closely than the modern use of radio waves is confined to the radio-telegraphy of Marconi's day. In 1981 only one American cable television system, in lower Manhattan, was doing much business carrying data, and its revenues from that service came to only $1.6 million. Yet various business forecasters are offering estimates of cable revenues from nonentertainment services of billions of dollars by the end of the decade. From being a system close to or even ancillary to broadcasting, cable becomes a system better described as a multiservice carrier.

New cable systems are being built with 54 channels on a cable and thus sometimes with 108 channels in total. Old cable systems are being upgraded to modern standards so that the cablecaster can offer profitable syndicated pay services.[31] But nobody needs 108 channels of conventional television entertainment shows. Twenty of these at once should saturate demand. There are a variety of uses for the other channels. Cities are asking for a few channels for education and municipal services. They are also mandating access channels to be used by community groups. All of these uses together may require a half-dozen channels. Then there is specialized programing of interest to small segments of the audience who are willing to pay for it, such as news channels, sports channels, children's channels, pornography channels, and high culture channels. Most of that material today, as well as advertiser-supported cable programing, is transmitted to local cable systems via satellite by national syndicators. There are now over 35 such syndicated services among which the cablecaster can choose. When penetration rates get high enough, similar local services will undoubtedly also emerge. Quite a few audio channels may also be fitted into the bandwidth of a couple of video channels. All of these uses together may soon bring the number of channels wanted up to about 40 or 50.

In addition, broadband cable systems can be used for teleconferencing, for business data communications, and for the delivery of electronic publishing services like videotex. Fifty-four channels is a fairly tight squeeze for what may be demanded; that many channels could be fully occupied well before the end of the fifteen years for which a cable system is usually built. Cities that insisted on twin cable systems have been wise.

Cablecasters themselves may like a tight squeeze. There are monopoly profits to be earned by not being forced to lease channels to competitors. Cablecasters will be happy to lease channels for purposes that do not compete with the pay services they themselves offer. They will be happy, for example, to link the branches of a bank on a data channel, or to lease a channel for a teleconference. But when a rival entrepreneur comes to a cablecaster who is offering pay sports or movie services and asks to lease channels so as to offer competing sports or movie services, the cablecaster may refuse, or if the franchise requires the offering of channels for lease, quote an astronomical price.

As cable systems become important carriers not only for entertainment but also for business, education, security, and public affairs, the issue of monopoly control will become increasingly acute. A cable system is a mixture of pluralism and monopoly. It has elements of each. It has numerous channels that can be programed by many separate producers. Video production is an intensely competitive business. However, one element of the cable system is a bottleneck monopoly, namely the physical cable. Utility franchises in the United States are almost always nonexclusive in theory. Thus, if a city government wished to do so, it could give a second company a cable franchise to operate in competition with the first.[32] But in practice, cable systems, like telephone and electric systems, are virtually always *de facto* monopolies; no second franchise is issued. Thus the natural legal analogy for the physical element of a cable system is neither the printing plant nor the broadcasting station, both of which are competitive, but the telephone common carrier system, which is obliged to carry whatever anyone wants to put on it at nondiscriminatory rates.

The central dilemma of cable is that it has unlimited capacity to accommodate as much diversity and as many publishers as print, yet all of the producers and publishers use the same physical plant. Just as all printed publications used the post office, so all cablecasts must use the cable. If the cable system is itself a publisher, it may restrict the circumstances under which it allows others also to use its system. However, if the cablecaster is not allowed to be a publisher but must be just a common carrier, the economics of the business is such that cable systems will probably not get built. The conundrum of how to have multiple competing publishers on a profitable cable system is what must be resolved.

When in 1970 a commission on cable television established by the Sloan Foundation prepared its report, some members and consultants favored a common carrier system. The commission majority, however, felt that such a system would not provide enough economic incentive to drive rapid and early investment into cabling.[33] Only those cablecasters who could earn profits from programing would have enough motivation to invest. Advertisers and others eager to pay to publish material via a common carrier would come along only when there was already a large cable audience. Thus, while ultimately cable should perhaps become a common carrier, for the moment, the Sloan Report stated, it had to be developed by entrepreneurs who would also be programers.

This pessimism about starting cable as a common carrier service could find support in the history of the press and broadcasting. The entrepreneurs who successfully started the first newspapers were often postmasters. When AT&T tried to operate broadcasting as a common carrier, it quickly found that WEAF had to assume the initiative in programing because no one else would. In general, early developers of a new distribution technology may have to take the lead in finding uses for the facilities they have on sale.

In 1973 a Cabinet Committee on Cable Communications reached a substantially similar conclusion in the Whitehead Report.[34] It looked forward to a system in which "cable would function much like the Postal Service or more appropriately like the United Parcel Service or a trucking company that for a fee will take anyone's package—or, in the case of cable, anyone's television programming—and distribute it . . . to the people who wish to have it. The key point is that the distributors would not be in the business of providing the programming themselves, but would distribute everyone else's programs to viewers that wanted to see them." But entrepreneurs in a new communications industry must be allowed to earn money in whatever way they can during the early years when the system is struggling. The Whitehead solution was that when cable penetration reached 50 percent of homes, the system should change from one in which the cable franchisee did the programing to one in which it was a carrier.

Against the evolutionary view of both the Sloan and Whitehead reports, one can object that once a successful industry has been set up in one way, it is unrealistic to expect that it can be turned around, reorganized, deprived of its main source of revenue, and told to earn its living in some other way. The British government tried this experiment with the telephone early in the century. The post office had been licensing private entrepreneurs to develop phone systems, but the developers were notified that the system might be nationalized in 1912, as it was. Those private developers did not put the resources and energy into the system that they would have if they had seen an unlimited future ahead of them. As a result, the British phone system developed more slowly and less well than in the United States. Despite that experience, a British government commission proposed in 1977 in the Annan Report that the same approach be used for cable. Private entrepreneurs were to be encouraged to develop cable so as to allow for a variety of experiments, but they were to be instructed that later the post office would take over the systems.[35] Such a prospect would hardly entice entrepreneurs.

Thus the failure of the American government to structure cable as a common carrier was not just for lack of libertarian principles or imagination. There were good economic reasons in the early years for considering a common carrier approach to cable impractical. Furthermore, broadcasting was a natural model for the FCC regulators to copy; they had lived with that system and believed in it. The Communications Act of 1934 recognized two quite different kinds of media: telecommunication common carriers, which were assumed to be basically monopolistic and were to be not only licensed but also regulated in their rates, and broadcasters, which were to operate in a competitive fashion to the extent that spectrum was available and without rate regulation. Broadcasting regulation has been less intrusive in the economics of the business than common carrier legislation as practiced by the FCC, but far more intrusive in content. The broadcasting rules governing cross-ownership and limiting owners to five VHF-TV stations with no more than one in a community do deal with economics, but their purpose is to force pluralism in every community and not to regulate earnings. The carriers, because they seemed destined to be monopolies but were passive as far as content was concerned, were regulated economically. The information that passed over them, however, was left quite free. For carriers there is no fairness doctrine or right of reply.

Despite the monopolistic character of the cable plant, the broadcast model was the chief influence on the FCC's thinking about cable. It was clear to the FCC that if cable was to begin to carry programing other than that picked up off the air, it would not be, at least at first, a passive carrier. So at the start of the 1970s the FCC saw cable service as broadcasting and thus as something with content to be shaped. But the FCC could not ignore the differences. So it designed its rules for a hybrid system.[36]

What the FCC wanted was a system that was partly a carrier, partly a broadcaster, but not to be called either. In the early years of CATV, the FCC had seen the advantages of having some channels like common carriers: "The public interest would be served by encouraging CATV systems to operate as common carriers on some channels in order to afford an outlet for others to present programs of their own choosing, free from any control of the CATV operator as to content ... From a diversity standpoint, it seems beyond dispute that one party should not control the content of communications on so many channels into the home ... The public interest would be served by encouraging CATV systems to operate as common carriers on some channels."[37] Since the 1970s, however, the FCC has retreated from this position.

The industry has fought vigorously against all suggestions that cable be run as a common carrier. Nothing stirs its fury as intensely as the words "common carrier." Cablecasters prefer to describe themselves as broadcasters or publishers who have the exclusive right and responsibility for editorial control of the selection and balancing of programing. They disregard the fact that the cablecaster has not one but maybe a hundred channels and is the sole franchisee in the community. The industry is presently engaged in an intense lobbying effort to have Congress prohibit the regulation of cable as a common carrier by either the federal government or the cities.

The industry is short-sighted. It is tempted by quick profits rather than a permanently viable system. In the short run, large profits can be made from movies, sports, and entertainment offered by a monopoly that is created by control of the physical cable. In the long run, public action against such a monopoly is inevitable. The economics of entertainment production and distribution is complicated. Protected by copyright, each entertainment product is a monopoly, but one in intense monopolistic competition with thousands of others. The few that succeed in becoming hits earn large monopoly profits. However, to do so, they must have access to the most popular and convenient distribution channels. Thus, theater owners, or television networks, or cablecasters, to the extent that they are not in a perfectly competitive situation, do enjoy some market power and can demand part of that monopoly profit. In each distribution system, the degree of market power is limited by the existence of alternative distribution systems. But just as independent television stations or video cassettes are no full substitute for network broadcasting, so satellite broadcasting, network broadcasting, and pay television, each with limited channels, are no full substitute for a mature cable system. And so the cablecaster, having some degree of market power, will seek to favor those profitable entertainment products which will yield up part of their monopoly rents. The interest of the cablecaster is not in maximizing ease of access.

From a social point of view the promise of cable lies in the pluralism made possible by its unlimited number of channels. From the programer-cablecaster's point of view, this may be its horror. A program producer gains from limitations on competition that compel vast audiences, because of the lack of alternatives, to watch programs of moderate interest. But for society, the advantage of cable is that it can create for video the kind of diversity and choice that exists in print. There can be simultaneously on cable, as there is in print, good programs and bad, popular programs and sophisticated ones, children's programs and adult programs, majority programs and minority programs, educational programs and entertainment programs, local programs and cosmopolitan programs, specialized programs and mass programs. There is no fixed limit to the number of possible channels any more than there is to the number of journals. Only cost sets a practical limit, as it does with print.

There are no easy solutions to the dilemmas of cable. At the dawn of cable, systems will only be built if their builders can earn a substantial portion of the profits from pay programing. At the maturity of cable, it cannot in a free society be other than a carrier. The transition will not be smooth. A major issue for the 1980s and 1990s will be how to prevent cablecasters from seeking the advantages of becoming publishing monopolists in their communities, controlling both the conduit and its contents. The issue has not become salient yet, because cable is still nothing more than a marginally improved way of delivering television entertainment, available to only a minority of the population. Unless the cablecaster offers good value, the public can still watch the same sort of material over the air or else buy cassettes or disks. But as more and more material migrates off the air onto pay channels, and as cable becomes the delivery system for all sorts of local and community and nonentertainment services, it will become important that the monopolist of the conduit not have control over content. Conversely, the ever more attractive monopoly will not be yielded up easily.

In the long run, the American public will not tolerate monopoly abuse of such a medium of abundance. If cable becomes an increasingly important means of delivery and nonetheless acts with hubris, then resentment and protest against a power wielder that overcharges, keeps unfriendly interests from talking, and behaves with arrogance will generate demands for reform. At the minimum, reform will enforce nondiscrimination and compulsory leasing of channels. Or it may go further and by law divorce carrier from content altogether. Or it may inflict bureaucratic government regulation on cablecasters, their practices, and their charges. Or it may impose public ownership on cable systems. One way or another, there is likely to be a rebellion against any system that gives one electronic publisher the means to select and control what gets published on the cable.[38]

The issue of cablecasters' power is just beginning to arise. In 1981 the cable industry tried to put through a bill that would have declared cablecasters not to be common carriers and would have forbidden compulsory access and rate regulation by any level of government—federal, state, or local. Mayors protested, and the Senate voted it down, but similar proposals will keep on coming.[39] On the other side, Henry Geller, a public interest attorney, has petitioned the FCC to adopt a rule compelling cable systems to lease channels.

The cable industry argues that it has no monopoly character because over-the-air television, direct satellite broadcasting, and videodisks all offer competing sources of video entertainment, while the telephone company offers alternative data lines. The time may indeed come when a broadband telephone network offers head-on competition across the board, but for now the denial of monopoly power is ludicrous. Whatever alternative means of communication exist, nothing else can offer the equivalent of the multiservice broadband cable running past every house, enjoying the privilege of a municipal franchise to string its wires and dig the streets. One can imagine a railroad owner in the nineteenth century denying being a monopolist because anyone refused access to a train could use a horse and buggy.

Cable systems clearly enjoy market advantages. One sign of this has been the shift in payment arrangements between competitive syndicators of advertiser-supported services and cable systems. At first some of the satellite-delivered services charged cable systems a few cents per subscriber per month to carry their material. Now, with the vigorous competition among syndicators, they are with few exceptions not charging or are even paying the cable system a few cents a subscriber a month to carry their service. The market measures where the power lies.

The monopoly advantage that cable systems have is conveyed to them by the state. A cable system is a government franchise that allows one company to dig up the streets in order to put its fifty or one hundred channels in front of every home. Those who seek and are denied access to the channels are so denied not just because economics makes it too expensive for them to publish in competition with the cablecaster, but also because a government franchising body has chosen one licensee. A private person may refuse others access to his facilities under most circumstances, but government under the First Amendment may not give the means of speaking to its favorites and deny them to others. A strong First Amendment case against such restraint by state action can be made on behalf of those seeking to lease cable channels.[40]

The issue of access will move to center stage when a large proportion of the viewing homes are on cable systems. The systems will then be money machines for anyone who can control them. Marketeers, authors, actors, candidates for public office, and others who seek to gain access to that massive audience will increasingly see themselves as at the mercy of cable franchise holders: "Motion picture companies, magazine publishing companies, book publishing companies, show business entrepreneurs and television and radio broadcasters should have full opportunity to use those channels and to compete."[41] Publishers are powerful. They have the instruments to mobilize public opinion. And they need channels of distribution. Publishers will no doubt have conflicts of interest, since large publishing organizations own many cable systems and are engaged in joint ventures with others, but in the last analysis the advantage to having exclusive arrangements for distribution in a few cities will be more than offset for the major national publishers by their need to distribute their information bases and entertainment products universally throughout the national market. The national publishers, backed by city politicians who wish not to be vassals of cable franchisees, and backed also by the public opinion of citizens concerned about maintaining an open marketplace for ideas, will probably succeed in destroying any regime for cablecasting that gives the franchisee control of the content business.

The present system, under which the cable franchisee is largely in the entertainment business, will be unviable in the long run also for other reasons besides the hostility of competing publishers and opinion makers. There are uses for a broadband network besides news, public affairs, and entertainment. Banks and brokerage houses in lower Manhattan are linking their offices and branches by the cable system for the transfer of data. For this service a system not requiring the investment in switches may be very competitive with the phone system. Also the broad bandwidth of cable is well adapted to the bursty and occasionally very high rates of computer data transmission. So urban cable systems could have a substantial future in the data carrying business. Video teleconferencing too is an attractive application for broadband networks. So is telemarketing. Coaxial cable can provide a video display of the product being sold along with the two-way message capability to close the deal.

Cable systems run by people in the entertainment business, however, have so far been slow to seize upon these opportunities. Most cable companies lack the technical competence to adapt their systems to such applications. They do not have the research laboratories to develop the equipment that business customers require for data security, diagnostics, redundancy, and maintenance. They do not have the technical personnel for prompt and reliable repair. The systems are designed for only the most minimal two-way communication. Transmission is noisy; the error rate (the percentage of erroneous signals) is low enough not to spoil a video picture but not low enough to satisfy a computer. On a switched system, if a particularly high quality link is needed between two particular points, it can be had, even if the rest of the system is noisier. On a cable system every message passes the full length to or from the hub. So noise cascades on a cable system, requiring sophisticated technology to control it. But few engineers are running cable systems; as in broadcasting, the few who are there serve mainly in minor service departments.[42]

Cable companies are also unwilling to move into the common carrier mode that would be appropriate for the marketing of data services since they see money to be made more easily by controlling the distribution of shows. Institutions that wish to use the cable for their own data systems or programs would wish to have their own origination points, almost like second head-ends; current systems are not designed that way. So there may be a delay in the transformation of cable networks into multiservice common carriers. But in the end, just as publishers are likely to insist upon it, so are the other industries that want broadband service. Some decades of public controversy can be anticipated before the issue is resolved.

If technology were standing still, one might seek a resolution on the basis of the key elements already outlined: the difference in the organizational and economic requirements of early and mature cable systems; the conflicting precedents from publishing, common carriage, and broadcasting; and the different needs of publishing institutions, data communicators, cities, dissidents, and the cable industry itself. But one more piece must be placed in the jigsaw puzzle. The technology is changing. The problem in twenty years will be different from the problem today.

Alternatives in Broadband Service

Off in the distant future of the 1990s or more probably the twenty-first century another alternative for delivery of broadband services seems likely to arise as innovations in the telephone network enable it to compete with cable systems. The time may come when producers who are denied the lease of a channel by a cablecaster will be able to send their video program to the public's homes over a broadband telephone network. Today's phone system is incapable of providing this service, and today's regulations also bar the phone company from such business. But the technical limitation will surely change, and in turn the regulatory one will probably fall. At that point cablecasters will have lost their monopoly of pay video programing, and may perhaps lose their whole business if the phone system is cheaper.

Two key changes in telephone technology are involved in this transition: the development of end-to-end digital transmission and the use of optical fibers.[43] The resulting system is an advanced form of what is called an integrated digital network (IDN). The acronym ISDN (integrated services digital network) refers to an IDN coupled with the services offered on it by the phone company. As with cablecasters, phone companies see the main prospect of profits in the service, not the carriage, and so seek to be not only a carrier but also a service provider.

Phone systems are converting from analog to digital switching and to digital transmission for many reasons. Digital telephone switches, which are computers, are cheaper, smaller, and more reliable than the electromechanical switches they replace. Since they are programable, new and commercially attractive services, such as call forwarding or dialed conference calls, can be provided with them. Digital transmission, which is well adapted to multiplexing, also reduces noise and attenuation problems, so it is superior to analog transmission for data communications. As a result, the core of the system, the switches and the trunks between exchanges, are steadily being converted to digital equipment. But at every point where the older analog plant meets the new digital plant, a codec, or converter between analog and digital coding, has to be installed as an interface, and this is expensive. To avoid such expense, in the end the whole plant will be converted to digital right to the black phone. But replacing the loops that reach from the local exchange to the customers' phones is also very expensive. Twenty-nine percent of the $120 billion phone plant is in that loop. Modernizing it is less important for the phone system as a whole than digitizing the trunks. Early versions of the ISDN will use as much of the present local loop as possible. So it will be some time before every home will have a phone connection that is both digital and very broadband.

When eventually this happens, and the present local loops are upgraded and replaced, the broadband digital channel coming into the home can be multiplexed to bring in different streams at different data rates simultaneously. Subscribers can talk on the phone, have their utilities metered, watch a video picture on their television, and receive their electronic mail, all at once without interference. The loop is likely to be an optical fiber rather than a copper wire because the fiber has the needed bandwidth at lower cost.

This development may toll the bell for cable television, but perhaps not. Unless the cable industry moves fast and aggressively in the next few decades to become an effective carrier for multiple services, it may well be displaced. If, however, it does move effectively, its unswitched plant should compete well in some activities. Cable systems have two advantages. One is that they may be the first network already in the nation's homes with broadband capacity. By the time regulations barring phone companies from the cable television business are changed and the needed sums have been invested to make broadband digital local phone loops universal, it may be well into the twenty-first century. The 1982 consent decree cutting off the operating companies from AT&T may further slow the process down, for the operating companies, which provide the local service, will be the less affluent part of the phone business. Cable systems are now spreading rapidly, and very possibly by the time the phone companies are ready to deliver video to the home, there will be a competitive broadband cable or fiber connected to most homes.

At that confrontation point in the future the question will be which broadband delivery system is cheaper. For broadcast-type transmissions the cable television plant may well be. One reason is that it has no circuit switches, which are an important element in telephone system cost. Since lack of switches is a drawback for point-to-point communication, the cable system is likely not to be competitive with the phone company for the universal service of person-to-person conversation. But for the mass medium job of pouring a common message to thousands of receivers down the lines, a cable system may prove cheaper.

The technical alternatives are complicated. Traffic on cable systems can and undoubtedly will be switched in some ways. Compare the topography of a conventional phone system with the tree structure of a cable system (see figure). All traffic on a cable system has to go to the head end or at least to a hub and flow down from there to every branch below. Traffic destined only for some subscribers, such as a pay television service, has to be scrambled and then decrypted at the legitimate receivers' premises. Thus, on a cable system switching consists of selective unscrambling of the traffic that flows by everyone. For some broadband services, such as a pay sports or movie service, this may be an economic approach. For narrowband service, like a conversation between two customers, it may not be. The calculation gets more complicated when there are joint products. A bank that has a cable channel among its branches for sending data may find it economical to multiplex message-switched voice phone service onto that channel. So cable systems, if they are aggressive and technically sophisticated, may end up competing with phone companies. The precise outcome of who has what part of the business will depend on technology, regulation, and entrepreneurship. There will, however, emerge from the phone system a common carrier alternative to cable systems, which will lease channels to those who wish and thus limit the market power of cable systems.

Other alternatives for broadband transmission to the customer are direct satellite links, cellular radio, and rooftop microwaves. These are also potential competitors to cable systems, though in less comprehensive ways than is a broadband digital phone system be-

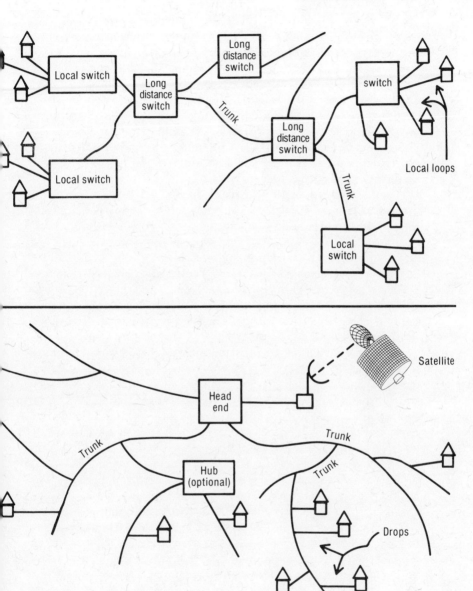

ircuit-switched Telephone Network (top) and Cable Television Network

cause, being over the air, they offer limited channels compared to enclosed carriers. They do, however, threaten cable systems in offering the few top entertainment programs, such as major sports events. Today, satellite broadcasting of high-rated shows worries cablecasters more than does telephone company competition. In a decade or so, satellite broadcasters may be able to deliver thirty or forty channels of national television programing either directly to homes that have a $100 to $300 antenna or to master antennas at apartment houses or neighborhood complexes. Technically this will be possible; economically it may not be. There may not be the market for thirty or forty national pay or advertiser-supported entertainment channels. So perhaps the direct satellite channels will be far fewer. Whether the number of such channels is four or forty, however, they will compete with cable systems for syndicated national entertainment programs. For many other purposes cable has distinct advantages. The social significance of cable is its numerous channels available for small local audiences. Over-the-air direct broadcasts will not compete well in this function. Satellites will not be the medium of choice for clubs, schools, and supermarkets to reach their local members and customers.

The various technologies of communications delivery—such as telephone, telex, videodisk, satellite, rooftop microwaves, cellular radio, ISDN, computer net, and cable system—each fits a niche in which it has distinct superiority, and it loses out in other uses. One or another technology is better adapted to local or long-distance delivery, switched or unswitched service, broadband or narrowband transmission, one-way or two-way communication, instantaneous communication or delayed communication, and universal interconnection or connection among a limited set of persons only. In the multidimensional matrix defined by these variables, cable systems are optimal for local, unswitched, broadband, one-way, instantaneous, universal networking. But if they are successfully established for use in that domain, they can also often compete in some other domains, earning better than marginal cost, by making modest technical adaptions. Cable systems can, for example, handle narrowband switched communication in competition with telex networks. Elementary economics will make cable entrepreneurs seek to levy differentiated rates, charging more for those functions where their technology enjoys great advantages, and less, as long as it is above marginal cost, for those functions where competition is severe. The competitors will, of course, do the same. Satellite companies, over-the-air broadcasters, and others will try, in addition to doing their own thing, to offer some of the services that are natural for cable at prices that will win them some of the business.

The result will be a complex market with cable systems having major advantages in some markets but facing stiff competition in others. And it will be a fluid market as the technology changes. No one can forecast well the picture for any particular moment of the future. Entrepreneurs must try to do so, for the rewards can be great, and so are the risks of error, but here it is enough to recognize some broad trends. The monopoly advantages that cable systems now have in multichannel, local, broadband service should become a more valuable advantage over the next five years as cable penetration grows. After a decade or more, however, the cable monopoly will begin to erode, particularly as a broadband digital phone system evolves. But for a formative fifteen to twenty-five year period, cable-casters will have a significant level of market control over one of the nation's major media of information distribution.

The primary advantage of cable is in narrowcasting. With an unlimited number of channels, groups of tens, or hundreds, or a few thousand persons located in a contiguous geographic area can join in receiving video, voice, and text at quite low costs.[44] Spectrum shortage problems inhibit over-the-air service for such small audiences. Cost problems preclude serving them over the present generation of switched networks. For cable to fulfill its promise, narrowcasting channels must be available for lease to those small groups who wish to use them.

Cablecasters are not hostile to this type of channel leasing. On the contrary, if they have idle channels, they gladly earn extra revenue by leases to small groups who do not compete with major entertainment activities. Education, high culture, church activities, club activities, ideological proselytizing, teleconferencing, electronic mail, and other such activities can pay better than marginal cost. A fourth of all systems with over thirty channels already offer channel time for lease.[45]

While cablecasters may be quite willing to lease channels for such community purposes, it is in their interest to charge such groups differentially what the traffic will bear. An hour for a course on Ikebana might be sold for very little indeed—anything above the marginal cost—but if the local Federation of Labor or Chamber of Commerce wants an hour they might find the costs substantially higher, and a movie distributor would find it far higher still. This is rational behavior for a cable entrepreneur.

The great bulk of the cost of a cable system is the sunk investment from constructing it. The marginal cost of putting a tape on an otherwise idle channel is minuscule. Charging an impecunious customer anything one can get above marginal cost may be rational, but unless from many customers the operator receives much more than that, bankruptcy for the cable system will follow. The viewers or the time leasers among them must somehow cover the average, not just the marginal cost.

The average cost of a cable hour varies with the number of subscribers per mile of a system, the number of channels on the cable, the hours of usage of those channels, and the miles in underground ducts versus on overhead poles. An hour on a small suburban system costs much less than an hour on a great metropolitan one. An hour on an old twelve-channel cable is more valuable than an hour on the fifty-fourth channel of a modern cable.

Yet generally the price at which cable systems with extra capacity can afford to lease channels is surprisingly low. An optimistic study in 1973 came up with an average cost per channel hour ranging from $4.30 for a standard twelve-channel system to $14.00 for an interactive one, on the assumption of a modest-sized suburban system with 50 percent penetration.[46] Inflation since then might double or treble those figures, or conversely, progress in technology might hold them down. In New York today, Manhattan Cable charges $37.50 for a half-hour of leased time and Group W charges $50.00.

There is no doubt that cable channels could be leased at prices that many groups and persons who want to be heard would be willing to pay. There are literally tens of thousands of commercial and nonprofit organizations that would find it worthwhile to lease channel time at such rates, even for tiny audiences. Schools might lease time for homework assistance. A labor union might lease time for a weekly meeting. A grocery chain or consumer group might lease an hour a night to quote prices and specials. An amateur theater group or orchestra might lease time for the satisfaction of having a real audience. High school teams might lease time to show their games. The city council or mayor might lease time for hearings or discussions. Churches might lease time for services. Entrepreneurs might lease time for programing for special interest groups. Anything exciting enough to elicit a collection from the audience of twenty or thirty dollars in response to the promoter's pitch could make its way onto the cable. So might membership clubs that could charge members a quarter a time. Besides these uses of cable to reach audiences, computer network operators would also lease time for their heavy traffic.

However, all those leases together, if charged at the basic rate for the bare channel, would not entice an entrepreneur to invest in a cable system. All the conceivable leases of time for other than mass media entertainment could not begin to pay for a cable system, at least in the near future. A cable system is a large and high-risk investment. Returns from the lease of channels alone would not get the systems built.

The major returns to a cable investor are from the subscribers' basic payments and from payments for pay entertainment services, such as Home Box Office and sports spectaculars. By May 1982, 13 million homes, or 16 percent of television homes, subscribed to pay services. Of these, only about 500,000 homes had the equipment allowing a computer at the head end to unscramble single shows for individually addressed subscribers, which the industry calls "pay per view." However, the addressable boxes at the viewers' sets are spreading rapidly, and soon millions of homes will be equipped to log in for individual events like $15 prize fights or $10 first-run movies. In addition, as the audience grows, ads, which in 1982 brought syndicated cable services only four percent of the revenue that they brought the three television networks, will be a growing source of income.

These are the cash flows that the franchise holders will seek to protect. They will oppose anyone leasing a channel at bare channel costs to put competing pay programing or advertiser-supported entertainment on it. Distributors would, if they could, bypass the franchisees and put programing of their own on a low-cost leased channel. In that way the monopoly rents that the franchisees count on to justify their investment would be transferred fully to the program producers or syndicators. The franchisees' indignation at being forced to lease channels to others who are ruining their business is understandable. Such a common carrier requirement in the early capital-intensive phase of cable system building would have stifled the industry's development.

Yet sooner or later it must become possible for anyone who would publish over cable to lease a channel, produce material as they wish, and charge the public for viewing it. If not, the market in ideas will be controlled by those privileged to hold a franchise to the cable.

It is not enough in the end to allow access programing, either free or leased, by community groups who cannot charge viewers for watching. Churches, schools, and governments may obtain the necessary funds to pay for time out of gifts or taxes. But a healthy, vigorous culture can exist only if producers can sell their product to consumers, as books and magazines are sold. Only thus can the return the producer gets reflect the value the consumer sees in the product.

Programs paid for by their consumers are generally superior and more diverse than those supported 100 percent by advertisers. How much advertisers will spend for a program is a function of its ability to sell their product, and that is only slightly dependent on the quality of the program. The number of persons who turn on a television set on a given evening is a function of habit far more than of the programs offered. With a small number of networks all aiming at the mainstream audience, the share that each gets also varies only moderately with the programing. The willingness of advertisers to spend large amounts to increase the pleasure of the viewer is quite limited. Since only a very small percentage of people who see a commercial will change what they buy the next time they are in a store, a spot commercial is worth only a fraction of a cent per person exposed. In fact, in the 1970s advertisers spent only about 2 cents per viewer hour to support television. At those levels most television production was "grade B" films. Economists concluded in the early 1970s that there was not enough advertising money around to support a fourth commercial network.[47] For the 1980s this conclusion is arguable, but regardless of whether the limit is three, four, or five, the point is that an advertising-based system has limits, set not by what the public wants but by what the advertisers are willing to spend. With that type of income structure, no commercial telecaster can afford to program for a small, specialized audience.

A pay system allows the audience itself to decide what will be offered. If people pay for specific programs that they wish to watch, their choices have less uniformity than in an ad-based system. People who pay for programs care about the specific content of what they buy. A program for which some people will pay $2.00 generates as much income with one percent of the audience as an advertiser-supported program earns at 2 cents a head.[48]

Among the design objectives for a mature cable system should therefore be that it has many channels, that much of the programing is paid for by the consumers on a pay-for-view basis, that all sorts of publishers and programers are able to put their material on the system on a paid or pay-for-view basis, that the cable plant owner does not discriminate among leasers by content, and that the owner is able to recapture for the physical system some of the large earnings of popular pay programing. Several features of a desirable organization of cable follow from these criteria. For one thing, cable operators should, as they are doing, install equipment to make customers individually addressable, to make possible pay-for-view programing and security monitoring. Upgrading the head-end computer, adding upstream bandwidth, and providing multiple computer and program access points will allow the cable to be used for business data communication, teleconferencing, electronic mail, and similar services. A large part of the future revenue of cable operators could come from addressing, billing, order taking, and data processing services rather than from raw channels. The cablecaster has a considerable advantage in doing that kind of processing on the head-end computer which is built into the heart of the cable network. Programers wishing to put on teletex or video material that goes only to physicians or only to those who agree to pay for watching a movie may develop their own facilities to do the addressing and billing functions with their own computers and set-top decoders and phone line reporting, but the cablecaster with the system's own computer is likely to be able to perform these functions more efficiently. So producers and publishers who lease channels to put on programs for pay will wish their lease to include use of that computer to clock and bill the customers.

A further implication of these considerations is that in charging for leased channels, cable operators should separate, or "unbundle," the rates for different parts of the service. That way they can segment the market so that various uses of the cable pay their fair share. Otherwise a radical separation of carrier from content might mean that one hour of a cable channel would cost the same for a church sisterhood as for a world championship prize fight. No principle of free speech or fairness compels such a radical conception of nondiscrimination. Nondiscriminatory rates need not mean equal rates for all services. Railroad tariffs were different for eggs and coal. Telephone rates are different for households and businesses. Postage rates vary with classes of mail. Nondiscrimination means equal tariffs for all customers seeking the same thing. It is not unreasonable to allow cablecasters to charge one rate barely above marginal cost for a high school play, a higher rate for a program with commercials, and a still higher rate for a channel leased for a pay showing of the championship prize fight.

A clean way to make such distinctions would be to establish a three-tier leasing tariff, in which one uniform charge would be levied for the raw channel regardless of what was carried, another charge for use of the billing computer, and a third charge in the form of a royalty on all revenue gained from advertiser or viewer payments for a program. Such purely commercial distinctions pass First Amendment muster. Any would-be publisher who wished to use the cable system would pay a publicly announced, rational, and content-independent tariff, but would end up paying far more for time used to peddle paid entertainment than for time for community chatter or for promoting ideas.

Thus the conundrum of how to make cable systems profitable enough to be built while at the same time giving equal access to all would-be publishers is solved. It is not necessary just to wait for technology to solve the problem, as it may ultimately do. To wait for a common carrier broadband ISDN to overwhelm the cable monopoly means that for a troubled decade or two of cable system dominance, the temporary monopolists would be in a position to establish a franchise on various parts of the video and electronic publishing industries. And many of this nation's leading and most effective publishers could be squeezed out of the market. It should not be necessary to tolerate a few decades of such market dominance.

An equally flawed panacea is to separate carrier from content now. The two businesses should be divorced, say some evangelists, and cable carriers should do no programing. That is both economically unfeasible and also wrong on First Amendment grounds. The American Constitution nowhere denies a carrier the same right as anyone else to publish without license or control. The Bill of Rights envisions no discriminations among classes of publishers. A cablecaster's freedom of speech is as broad as that of those who lease his lines.

The core problem in such a situation is how to ensure that cablecasters do not discriminate so as to favor their own communications. A cablecaster may be required by law or franchise to lease channels to all comers under a publicized tariff structure. But if the tariff for leasing is not regulated, the carrier may set rates so high that competing organizations will not choose to lease time. For the carrier to pay those rates to itself is only a bookkeeping matter.

An ingenious structural device to achieve competitive market in cable access was invented for the city of Boston. Its cable franchise allocated some channels to a Public Access and Programming Foundation. The foundation may lease channel time to others. As a result, the franchise holder itself has an incentive to lease channels at a reasonable fee, for if it refuses, the applicant may go to the foundation and lease a channel there, getting the same thing with no revenue to the franchise holder. There is, in short, a duopoly in suppliers of channels instead of a monopoly.

Insofar as cable franchises turn out in practice to be an opportunity for making excessive monopoly profits, cities may be tempted to regulate rates, as they do for other public utilities. However, experience suggests that even as an economic matter, there are better ways of attacking the problem than traditional rate-of-return regulation. One device for ensuring reasonable tariffs for leased channels without subjecting cablecasters and others to the bureaucratic burdens of rate-of-return regulation is arbitration. A franchise can specify that unused channel time must be made available for lease under a tariff card—perhaps a three-tiered one—such that for the physical facilities the leasers pay their proportionate share of physical plant, operations, taxes, capital costs, and a reasonable profit—in short, all costs but programing. While allowing any posted pattern of variation by time of day or other such parameters, the cable operator's charges for the raw cable and head end computer can be limited to some standard of reasonableness of the net result. Such tariffs can be written without regulation, but if the enfranchising authority concludes that rates do not reflect the principles laid out in the franchise, it could be authorized to demand arbitration. In most arbitrations each party chooses one arbitrator, and the two of them choose a mutually agreeable third. This is a laxer scheme than having a bureaucracy regulate the rates. The cablecaster might earn marginally more, but arbitration could insure against exorbitant rates, and do so without creating a permanent bureaucracy for enforcement.

On successive renewals of franchises, cities can gradually shift the terms away from the initial broadcasting conception of the cable system to a common carrier conception. As the volume of business grows, the common carrier approach becomes more appropriate. Since no franchisee is guaranteed renewal of a franchise, the entrepreneur from the start has to calculate a budget to recover costs within the franchise period, which generally runs fifteen years. No confiscation would follow from obliging systems to lease channels more liberally under successive franchise renewals. For instance, when a city's cable system has come to be pervasive and heavily relied on, renewal franchises might well expect it to evolve from a tree with a single head end to a distributed topology with many sources of origination.

Most important of all, cities should require large numbers of channels on the system. If no idle channels are left at the busiest times, then the system has been too conservatively built; there should always be some. When channels are idle it is in the cablecasters' interest to lease them, even if not in ways that compete with their own top entertainment offerings. And when many channels have already been leased, more rational terms can be calculated even for those leases that the cablecasters might prefer to avoid.

The main responsibility for ensuring free and pluralistic cable networks that allow leased access for all who wish it lies with the cities. They issue the franchises that give cablecasters their monopoly rights. They specify the access and tariff requirements. The First Amendment places limits on what cities may do, but within that scope they may set up their cable systems in a variety of ways. Some will move toward a pluralistic system of cable access faster and others more slowly, but the direction of movement for a free society is clear.

In the United States, the principles of common carriage have been applied fairly well in the case of the postal service and the telephone system. Although there have been issues, those systems now give the carrier very little control over what it carries for the public. With the newest common carrier, cable television, there is danger that everything will be done wrong, as if nothing had been learned. The system that has been created merges monopoly with publishing. It will take enormous political wisdom for the country to work its way through to a system in which what is carried on the physical plant is, as in the case of print, completely uncontrolled, and which, during the twenty or thirty years until broadband telephone competition becomes a reality, will give to all those who make the effort to produce and publish, fair access to that monopolistic network of channels which they need if they are to express themselves via cable.

8. Electronic Publishing

If computers become the printing presses of the twenty-first century, will judges and legislators recognize them for what they are? At issue is the future of publishing. The law till now may have infringed the First Amendment for newer media, but for the printed word, freedom from licensing, from prior restraint, from taxation, and from regulation has remained sacrosanct. When written texts are published electronically, that may not remain true. The law on the use of print, the mails, telecommunications, and broadcasting is by now cast in a mold of precedents that is sometimes libertarian, sometimes less so. But for still newer technologies where the law is in gestation, precedents are in the process of being set, often in ways to cause alarm. Practices are now being canonized in regard to cable television, computer networks, and satellites which may someday turn out to be directly relevant to publishing. People then may ask in puzzlement where protections of the free press have gone.

Calamity is not foreordained. The trends in freedom have not been one-way. The technological trends are favorable, though the political response to them has been less so. Eternal vigilance there has not been; more often what has prevailed is unawareness that actions taken had First Amendment implications. Regulations are set for technical reasons or for the protection of established interests, and only in retrospect are they seen to impair free speech. Then in a burst of occasional vigilance a court may blow the whistle. Having previously sanctioned a path that led to abuse, judges then find a line of reasoning to distinguish the case at hand from those that led up to it. This has been the pattern so far. There is reason for concern about where it will leave freedom as the publishing of books, pamphlets, magazines, and newspapers becomes electronic.

Paper and Screen

Publishing is becoming electronic for reasons of both convenience and cost. Large, complicated arrays of bits of information can, using computer logic, be edited, stored, transmitted, and searched with a flexibility that is impossible for ink records on paper. Millions of words can be scanned in seconds and transmitted across the world in minutes. Up to now, however, this convenience was bought at a price. Electronic text handling was powerful but expensive; paper records were cheaper. That situation is reversing. It is becoming cheaper to handle words electronically than to handle them physically, to the point where the physical mode is becoming too expensive for ordinary use.

People are not going to stop reading from and writing on paper, or carrying pieces of paper in their pockets and purses. Nothing on the horizon is yet the full functional equivalent of this most useful technology. People did not stop talking when they learned to write, and they did not abandon pens when typewriters were invented. Use of paper is likewise not likely to be given up for any alternative that can be anticipated today. Cathode ray tubes (CRTs), television screens, or microfiche readers are often less comfortable to handle than printed sheets. Yet increasingly the most economical way of moving, storing, and displaying words is electronically. The use of paper is becoming a luxury.

Of four distinguishable functions in the processing of words—input, storage, output, and delivery—storage is already cheaper in computers than in filing cabinets. At the moment, the most promising storage technology is that of videodisks. An optical digital six-disk pack is anticipated to appear shortly with a total capacity of a trillion bits or over a hundred thousand books. At an estimated cost of $51,000 for this device, a book as a stored object—not including its editorial, composition, or distribution costs—would cost forty cents. One hundred such videodisk packs, or the equivalent of the Library of Congress, would cost $5 million and fit in a medium-sized room. Other longer-term projections for different videodisk approaches arrive at cost estimates of a cent to a nickel per book.[1] Clearly the significant costs will not be those of storing the data.

To transmit material electronically is also far cheaper than to print it out and carry it by plane or truck. Compare the Hewlett-Packard Company's interplant electronic message system's cost of about a penny per hundred-word average message for domestic transmission with the cost of a first-class stamp.[2] For storage and delivery the choice is clear: computer memories and networks far outdistance paper as the most economical medium. The balance between paper and computer is less one-sided for input and output. Higher than the cost of electronic storage and transmission is the cost of getting a microscopic record that is on tape, disk, film, or bubble memory back out again to someone who wants it in a form in which it can be read, and still higher is the cost of input.

In costs for retrieval, electronic means are gaining the advantage. The cost of labor for finding one sheet of paper in a file containing millions of sheets is rising, while the cost of computing hardware is falling by a factor of about 40 percent a year. There is little doubt that searching for a record in computer memory and displaying it on a screen, or even printing it out again on a sheet of paper that can then be thrown away because the permanent record is still in the computer, will become, even if it has not yet become, cheaper than getting it out manually.

The main continuing use of paper in a mature electronic communication system is likely to be as a medium for active handling and short-run retention, and this use may grow enormously. The probable style of such usage is illustrated by a Xerox laser printer, which is both a copier and a computer on a network. To retrieve a manuscript from a file in the memory of any other computer on the network, someone calls the document up, specifies the format, type fonts, and other visual details, and the laser printer instantly runs off one copy or as many as are immediately needed. The important difference from a 1970s photocopy machine is that no paper original has to exist to be copied.

The paperless office or paperless society is probably a fantasy. Though for both storage and transmission, paper is likely to become a rarity because of its cost, the use of paper for display, reading, and current work may grow, partly in fact because it will not be economical to retain the paper copies. The paper industry has cause for optimism. Experience shows that when word processors are introduced into offices, paper consumption increases, since with a word processor it is easier, when minor corrections are made, to run a whole new version of a document rather than tediously to correct the old copies. So too, when no paper files are kept because bulk storage of them is too expensive, a new paper copy may be derived from the bulk electronic files every time an item needs to be seen, and then that copy can be thrown away.

These points apply not only to office records and memos but equally to books, articles, and reference materials. There is no reason to stockpile printed paper when a laser printer can produce a paper copy whenever a customer wants one. Even in the 1960s farsighted library planners looked forward to the day when people would never take out the library's unique copy of a work but would instead get their own instantly printed copy.[3]

This may seem wasteful of both forests and money. It is, and paper will be used only because it is the most convenient if rarely the cheapest way to display text. A screen display, if acceptable, is generally less costly. Just how much cheaper is a question that will vary with time, depending on electronic and paper technology, both of which are changing. Recycled paper, erasable ink, synthetic paper, and cultivated cellulose instead of trees could all affect the calculations. But nothing on the horizon suggests that physical storage and transmission, or even retrieval, will be able to compete with the electronic alternatives.

The balance of costs between paper and computer is least one-sided with regard to input. Off in the dreamland of inventors is an optical character reader (OCR) that can look at any font of type or any handwriting and turn it correctly into digital representation, but such a reader does not yet exist. Success has been achieved cheaply and reliably only with well-defined fonts. Without such a device, the entry of text into electronic form requires the manual operation of retyping, and this is horrendously expensive. Without an optical reader or its equivalent, the Library of Congress will never be put into computer memory. No one will pay the cost of typing it all.[4]

In short, to put records into computer readable form is rarely economically justified if the purpose for doing it is solely to file them. However, if for other reasons the records need to be put into computer-readable form, then it becomes more economical to file and retrieve them that way than to print them out and preserve them in hard-copy form. Since more and more of everything that gets written does at some time appear in computer representation, its further handling in electronic form is dictated by economics. In the near future, virtually everything published in print will surely be typeset by computer. So too with the coming of word processors, whatever is done by the present 4.8 million stenographers, typists, and secretaries in America will exist in electronic form and be accessible for computer storage, transmission, formating, and editing. Thus economics dictates the fact that most text handling will become electronic.

On-Demand Publishing

The very definition of "publishing" is changed by convergence between books, journals, and newspapers, which deliver information in multiple copies, and information services, office automation, and electronic mail, which deliver information in editions of one. A distinction between publishing and the provision of individual information was a product of Gutenberg's mass media revolution, when for the first time in history written texts could be mass produced cheaply. With contemporary communications technologies, singly produced copies are no longer much more expensive than mass-produced ones.

The Xerox copier was a major breakthrough in reducing the economic advantage of mass printing. For people who do not put a high price on their own time and who are insensitive to the law, making a copy of a printed publication today is highly competitive with buying one in a store, which has to bear the costs of inventory and distribution. Computer processing of text is pushing the process further. Data communications, a point-to-point medium, is falling, as we saw in the figure in Chapter 5, toward cost levels like those of print mass media and is rising correspondingly in volume of use. In electronic publishing it is not necessarily economical to preprint multiple copies and physically store and distribute them. It can be cheaper for the source text to exist in some microdigital form, to be published later in units of one on reader demand.

Such on-demand publishing is increasingly used for technical literature. It is estimated that of the 2.5 billion copies of journal articles published each year in America, 250 million, or only one in ten, are ever read.[5] Providing each reader with selected sets of articles reflecting an individual profile of the reader's interests rather than sending entire, uniform journals has clear advantages. Services of this sort are growing.

Electronic publishing and office information processing are both being revolutionized by the same technologies, so they tend to converge, even though a difference in purpose remains. Publishing remains a business devoted to organizing texts and information and offering this material widely, either in the marketplace or for free. Whether a publisher preprints and warehouses multiple copies or prints on demand is a matter of market strategy, not of cultural substance. Publishing's strategies may change, while it retains in an electronic era as vital a function as ever. It remains the vascular system of science, democracy, and culture. The cause for fear is that when its technology looks like that of an office the law may see it as commerce, not publishing, and thus subject to regulation like any business.

New Technologies

The technologies that are transforming publishing include the computer, reprographics, data networks, and electronic storage. Some of the media are so novel that we still grope for generic names for them. What now are referred to as data networks, computer networks, value-added networks, videotext, videotex, teletex, teletext, or viewdata may at some time in the future meld into an integrated network of digital services. But today there is no accepted name for the class. Because the word "network" is likely to survive to describe the electronic transmission of signals among numerous points, it is used here as the generic label.

Geographically distributed electronic storage devices, too, have as yet no generic name. Today they include mostly magnetic disks and tapes. In the future these forms may be replaced or joined by bubble memories, or videodisks, or Josephson junctions supercooled to temperatures near to −460 degrees Fahrenheit, or something as yet unknown.[6] Since at the moment the most promising portable device for storing large volumes of text is the videodisk, the word "disk" is used here as shorthand for magnetic tape, videodisk, or any other device for distributed mass storage that may succeed them.

While networks and mass storage devices are the main technologies for the transmission and retention under computer control of electronically published information, there must also be devices to translate the data to a form that human beings can read or hear. These output technologies either print words and pictures on paper, display them on a screen, or reproduce speech. In the primitive technologies of the past, methods for transmission, storage, and display were closely linked. Text or pictures stored on paper were displayed on paper. For words to come out of a loudspeaker, words had to be spoken into a microphone. This is decreasingly true. What is stored in digital memory can be put out in whatever form is most convenient. The choice to have the output oral or visual, on paper or screen, is independent of the mode chosen for transmission or storage; it is a choice based on purpose and cost.

Symbols digitally encoded within a computer can be disseminated in many different ways, leaving it to publishers to calculate trade-offs, such as between storing information centrally for transmission as wanted and transmitting it once to the customers' disk, there to await the need to be seen. The optimal choice depends on usage patterns for the type of information. Some information is almost eternal; some of it is good for long periods; and some is ephemeral and needs constant updating. The proportion of each type of information varies with both the use and the user, as shown here:

Information durability	Information user				
	High school student	*Stock trader*	*Theater goer*	*Historian*	*Engineer*
Eternal	History Classics Math Dictionary	Market records Corporate records	Classics, plots and authors	History: Who What Where	Mathematical formulas Physical constants
Long-lived	Pedagogical exercises Novels Civics Science	Regulations and laws Companies, their markets & structures Who's who	Modern plays Stars	Recent scholarship: scholars, theories, interpretations	Standards Codes Companies Product catalogues
Ephemeral	Current events Class schedule Homework assignment	Current market quotations Business news	Theater schedules	Current publications Journals	Changes in: standards, codes, companies, product catalogues, RFP's, contracts

For ephemeral material the user needs a fast delivery vehicle, be it the morning paper thrown on the front porch or a computer network tapped by a terminal. For eternal material the economics probably dictates that it be stored at the user's home or office in a book or on a disk or similar device.

In the 1960s computer economics seemed to be dominated by Grosch's law that the cost efficiency of a computer grows with the square of its size. It followed that computers would end up being public utilities. Everyone would make use of one or a few giant computers in time-sharing mode. Driven by economics, most information would ultimately be stored in a vast central archive, with distant users gaining access by way of a network. This prospect struck fear into Congress and the public. A horrendous future loomed, in which life records on 200 million Americans would be stored in one archive, with social security, internal revenue, military service, education, and credit files all linkable in one place. In alarm, Congress voted down even modest proposals for computer data archives.

Now in the microelectronic era it seems clear that Grosch's law was wrong. No general economic advantage accrues from the centralization of archives or of computing centers. The economics works the other way. Microcomputers are proving more economical than big ones for most purposes. Telecommunication costs are falling more slowly than computing costs. It therefore pays to move stores of stable information only once to a user's premises and to use a microcomputer there to search, retrieve, and organize the files.

How most economically to handle a file of long-lived but not eternal data depends on how often it needs to be retrieved, what transmitting it by telecommunications costs, what memory costs at the user's location, and what processing costs locally and centrally. It may pay to distribute some files to each user on local disk-type storage, or for other files, it may pay to store them centrally and gain access to them by telecommunications when needed. The best solution varies not only with the user and kind of data but above all with changes in technology.

Since every user works with some mix of long-lasting and fugitive information, an electronic publishing system will be a mix of local storage devices on the users' premises linked by networks to publishers of ephemeral data and to publishers who update long-lasting data and maintain archives, from whom the users may obtain additional data from time to time. Some of the disks may be carried to the users' premises by mail or by the users themselves after being bought at a "bookstore." But it will often prove cheaper and easier to deliver the information over the telecommunications network, to be copied on the users' premises by their own recording machines. If one wishes to gain access to data in a data library, say Lockheed's in California, one can have it transmitted over a network, such as Telenet or Tymnet, directly into one's own computer—doing it late at night if one needs to keep communications costs down. However, for large files the local storage medium that people today are likely to possess, whether magnetic tape or disk, is fairly inconvenient or expensive. Videodisks read by lasers are cheap and convenient, but at present they have to be published, like books, in capital-intensive factories. With the appearance of predicted systems for the local writing of data by telecommunications onto videodisks, the physical shipping of disks may remain competitive for large editions, but much data will be read onto the users' own disks via networks. *Jaws* may still be sold on disks at computer stores, but a collection of journal articles needed by a physician treating a particular disease will more likely come by telecommunication. Thus, disks and networks are alternatives that will grow in tandem.

First Amendment issues are likely to become more acute, and confusion about them greater, the larger the role of networks in electronic publishing and the smaller the role of mass "printing" of videodisks to be delivered by hand. The analogy of videodisk publishing to print publishing is so close that courts are likely to see them in the same way. The fact that a book is encoded in a square centimeter of the surface of a videodisk is not likely to make much difference to a judge's perception of it.[7] If the disk is produced by a publishing company in a few thousand copies, sent to bookstores, and sold there to customers, a judge will probably see it as akin to a book. It not only is publishing but looks like publishing.

Information retrieval over a telecommunications network, however, looks like something quite different. It too is publishing, but it looks less like the printing and publishing industry that the courts have surrounded by First Amendment protections. Electronic publishing may start by using computers to bypass the costs of union composition of physical end products that are conventional newspapers and books. But at later stages it may look much more like a telephone or cable system, which the courts are used to regulating, even though it incidentally but importantly prints out words.

To retrieve material from a computer, a reader types instructions at a terminal, which answers by displaying messages responsive to the instructions. The information contained in the messages may

enter the computer as mail from some other person, or as a file arising from administrative operations, or as an organized and edited
data base placed there by a publisher, or as a posting on an electronic
bulletin board where members of the community place material that
others might wish to see. From a technical point of view, all these operations create data bases for information retrieval, and the method
of retrieval is essentially the same. The lines between these operations are fuzzy and will become more so.

In the United States by 1982 there were about 1200 publicly
available on-line data bases created by some 700 organizations, the
number of data bases having tripled in three years. In 1980 American
libraries of all sorts conducted 3.9 million on-line searches of computerized bibliographic data bases. Bulletin boards and information
exchanges among interest groups are proliferating. There are 70 of
them on one network of computers (Unix systems) and 50 on another network (the Arpanet), covering such subjects as cars, aviation,
space exploration, recipes, motorcycles, games, jobs wanted, jokes,
limericks, math, movies, office automation, taxes, travel, wines, and
even computer bulletin boards themselves. An arrangement called
the Amateur Press Association allows aspiring writers to put their
output on computer networks for their colleagues to read. Thus, information retrieval may not necessarily produce multiple copies of
any physical object that reminds one of a book or magazine, and yet
it publishes a flow of opinion and knowledge.[8]

The electronic networks that are now emerging alongside telephone, radio, and television, unlike those older electronic devices,
are not just parallel complements to the print media; they are actually transforming print. Telephony never became a one-to-many
mass medium. Broadcasting, although it competed with the press,
remained a separate industry. Both of those electronic industries
operated in different markets and in a different legal environment
from print publishing. They changed the First Amendment regime of
the press little, if at all.

This may not prove true for electronic storage and data networks.
Disks and computer networks, like the press, deliver written texts.
Networks may use telephone lines, radio waves, or coaxial cables as
physical links, but whatever their transmission medium, they do
carry written news, mail, and information services. They constitute a
new electronic form of publishing. They converge directly with the
printed press.

Printed information delivered electronically may sometimes be in content and appearance exactly like the old. For example, the *Wall Street Journal* is sent as a facsimile by satellite from a dish at the production headquarters in Chicopee, Massachusetts, to dishes at local printing plants around the country, where plates are made and a conventional paper run off to be delivered to the customers nearby. Alternatively, data networks may carry identical information to that in the print media but present it in different formats. For example, the same Reuters stock market tables which one person reads in a newspaper another customer reads off a CRT. Or a totally new form of electronic information delivery may displace the old media. For example, a customer at a computer terminal may program Boolean searches to retrieve a fact that other persons still find by reading through the pages of reference books. Or if the goal is poetry rather than facts, one person may read a little magazine, while another feeds key words and cadences to a poetry-writing computer program. Whatever the precise situation, electronic information systems are in the same business as the print media and will therefore displace or converge with hard-copy publishing. It would confuse the mechanism with the function to subject data networks and storage devices to legal precedents from the previous electronic media rather than to the law of print, and the consequences would be dire.

Networks

Networks, like Russian dolls, can be nested within each other. In current regulations a distinction is made between the underlying physical carrier and the service networks that operate on top of it. For example, AT&T Long Lines operates a nationwide microwave network of physical towers with energy radiating between them. At the next level a subsidiary company, like American Bell, or a competing company, like GTE-Telenet, leases circuits from AT&T on which to operate a network linking some set of computers and terminals. A third-level network can be placed on top of this network to link a group of subscribers to a particular information or message service on those computers. Those linked can be travel agents, or philatelists, or members of a religious sect, or dealers in modern art, or any other group that has reason to communicate. So there can be networks within networks within networks.

Among the basic physical carriers are not only telephone networks but also cable systems, satellites with their ground stations, private microwave links, and mobile radio-telephone systems. Some of these, called specialized common carriers, compete directly with the phone company. The first specialized common carrier was MCI, which in 1963 applied to the FCC for the right to build microwave repeaters from Chicago to St. Louis to provide a mobile telephone service primarily for the use of truckers on that route. This request challenged AT&T at the heart of its business: long-distance voice service. AT&T argued that MCI would serve only such profitable routes, skimming off the cream and leaving AT&T with the obligation to serve the less profitable routes. The FCC nevertheless granted MCI's application in the belief that it would encourage competition. Since then, numerous similar applications have been granted. For example, the Associated Press has a network of several hundred earth stations at broadcasting stations and newspapers that receive AP news from a satellite.

The distinction is blurry between the underlying physical network and the service network immediately on top of it. There are functions that either of them may provide. For example, routing and directory service can be viewed as part of the basic service of getting the message from place to place or as part of the enhanced service that someone other than the underlying carrier can also provide. The temporary storage of a message when the intended receiving terminal is busy or out of order is another borderline activity. So too is converting the format when the sending and receiving terminals operate at different speeds or with different codes. But fuzzy and disputed as the distinction may be, a border exists. At any given time some basic transmission operations are performed by the organization that provides the physical plant, while additional functions are provided by other organizations that lease its circuits and then sell service.

A conflict exists between the interests of the underlying carriers and the entrepreneurs who lease circuits from them, and for that matter between the interests of the operators of the second-level networks and their third-level customers, such as publishers. The entrepreneurs at the more basic level prefer to incorporate as many salable functions as possible into their service, while their customers prefer to bring as many functions as possible up into their own business; at issue is who earns the revenue from each function. Cable operators, for example, seek to expand their functions upward into programing. AT&T presses the FCC to designate more functions as basic rather than enhanced services. American Bell's Advanced Information Service (AIS) is a network which differs from previous data communications offerings most markedly in that it proposes to offer all sorts of specialized data-processing services to various industries, such as travel agents, at the nodes of the network itself rather than leaving those tasks to the customers.

In a competitive market this conflict of interests does no harm. Vendors offer various services; customers who do not wish all of them can seek a vendor who sells a stripped-down version. If travel agents just want a line to the airlines and hotels and want to do the reservation processing, billing, and accounting on their own computers, someone will, in a competitive market, sell them that service. If some publishers want to connect information bases to customers by networks that incorporate formating functions, while other publishers wish to do the format programing themselves and lease only pure communication lines, in a competitive market they can.

The underlying network, however, is often a monopoly, at least in part. One reason is that the physical plant may have to cut through streets or private property and so may require franchises and the right of eminent domain. Another reason is the convenience of having the physical network "universal" in reaching every house and office. There is usually, therefore, only one set of telephone wires or video cables through the streets of a city. Common carrier principles are designed for such monopoly facilities; the public needs access to the plant. The situation regarding service networks that use the underlying facility, though occasionally similar, is more often quite different. So long as nondiscriminatory access to the underlying network is a matter of right, then any service vendor can utilize the physical network. Such service providers will usually find themselves competing with others, for their activity can be duplicated many times over on the same transmission base.

Value-Added Networks

The term commonly used for a service network that operates on top of the physical network is a "value-added network," or VAN. To serve its customers, such a network uses lines that it leases from the phone company, from a satellite carrier, from a specialized common carrier with microwave facilities of its own, from a cable system, or from a combination of these. The simplest example of a VAN is a communications broker or wholesaler. In the United States since 1976 it has been legal for any group of enterprises, such as airlines, to lease lines from the phone company and share their use, or even for an entrepreneur to lease a line and then resell its use to others who need less capacity than a whole private line of their own.[9]

It saves customers money to share high-speed circuits that have a greater capacity than they individually need, if the structure of tele-communications tariffs is such that the charge for transmitting data at a given speed goes up less rapidly than proportionately to the data-rate or speed.[10] If, for example, it costs substantially less to lease one broadband circuit on which 48 1200-bit/second circuits can be multiplexed than it does to lease 48 such lines separately, then a wholesaler can make money by leasing a broadband line and then subleasing it as 48 such circuits. A wholesaler may help custom-ers organize the shared use of broadband circuits. But to assume that the efficiency of such wholesaling is inherent in the technology of telecommunications because high-volume transmission is cheaper per bit than low-volume transmission is a common error. If a phone company is efficient, as the American phone company is, the whole-saler can do nothing in batching and concentrating the traffic of sev-eral customers that the phone company itself cannot also do. So the wholesaler's service is really to take advantage of irrationalities in tariffs.

In the United States, wholesalers of long-distance telephone ser-vice who today offer rates below those of AT&T include MCI, Sprint, and City-Call. One reason these wholesalers can undercut AT&T is that they serve only heavy-traffic routes. Another reason is that long-distance rates have historically been set too high and local rates too low. The Bell System has compensated for this discrepancy by settlement payments from Long Lines to the operating com-panies. Although things will change under the 1982 AT&T consent decree, until recently long-distance wholesalers shared no responsi-bility for local service. Thus resold circuits on high-speed leased lines were cheap.

Florists used this fact to go into telecommunications wholesaling. For decades flower stores with the Hermes logo on their windows have offered delivery in remote cities. Originally they ordered by telegraph; now they use telephone. The florists share a network of leased lines among major cities. They connect to that intercity network by a local call to a node in their own city. But the florists do not use the full capacity of the long-distance lines that they lease, so through newspaper ads they have offered service on their network to others. The nonflorists who subscribe use the network to call other cities, just as if they were florists. So the florists are in the telecommunications business as wholesalers.

A more interesting kind of value-added network allows something to be done other than just brokerage. On such a network, computers process messages in ways that add some substantive value besides transmission. For example, the network may route messages. Polling systems collect messages from and deliver them to terminals on a network several times a day. A central switching computer polls the terminals on a regular schedule and then stores and delivers the messages that it finds in them. The connection is made by calls over the public telephone network. These calls are dialed by the switching computer itself on a schedule of, say, every hour, twice during the business day, or once a day; any schedule can be set. The messages can be person-to-person letters. They can equally well be manuscripts, publicity, news reports, or other published material, addressed to one particular terminal on the net, to a set of subscribers, or to everyone on the net.

Another kind of now value-added network, offered by ITT, enables customers with facsimile machines of different brands to send facsimiles to each other. Such machines are often incompatible, but the ITT computer has stored in it a list of the protocols for each brand and model and also the brand and model that each subscriber owns. So a facsimile goes from the sender to the central computer, from which it is sent under the necessary protocol to the addressee's machine. Other common value-added networks store, edit, and reformat messages and assure compatibility between diverse terminals. Such a service can also include directory information or mailing lists so that a message goes to its target, say the resident hematologist in Podunk, without the sender having information on the name or address.

Sophisticated service networks that are superimposed on the underlying physical network are designed to meet the specific needs of computer communication. Computer traffic, in contrast to phone traffic, is extremely bursty. In both kinds of traffic there are periods of silence and periods of talk, but in phone conversation when a person talks, there is a moderately even volume of sound for at least several seconds in a row, and conversation goes on most of the time for several minutes between the same two people. When computers transmit, they can send a great burst of bits of information in a tiny fraction of a second, then be silent for a long time, and then send an entirely different addressee another large burst of material for another tiny time slice. A conventional telephone switching system that takes a few seconds to set up a call between persons, who then use their moderate-bandwidth circuit for a few minutes, is not economic for calls between computers, which send a burst of data requiring an expensive large-bandwidth circuit but only for a tiny fraction of a second for each connection.

So as not to waste the dead time on the large-bandwidth channel that is required to handle the occasional data bursts, computer networks are designed so that multiple users share the same channel, each taking occasional fractions of a second of time; this is called time-division multiplexing. To load the system efficiently, messages are often broken up into standard-length packets for transmission. A network that uses that form of multiplexing is called a packet network. A small computer at each entry port to the network breaks an outgoing message into packets and sends them on through the network to an interface computer near the destination. As in a cable system, all traffic flows intermingled together; no dedicated pair of wires is used for any message. Thus, much of the traffic that flows through a node is not destined for it. An address on each packet is read by each node as the packet passes it; those addressed to the node, and only those, are picked off. The receiving interface computer then reassembles the original message out of the packets into which it was divided.

The advantages of a packet network are several. First, it can connect various kinds of incompatible computers, since the interface computer through which each operating computer connects with the network takes care of the conversion of messages to and from the format used by the network. Second, the packet technique enables many users to share the expensive high-speed lines and keeps the lines evenly loaded. Packet techniques can be used for voice and other kinds of traffic but are now mostly used for data and message traffic. Among the most important packet nets are the airline reservation networks, the bank clearance and payment networks, and the public networks like Telenet, Tymnet, and AIS which offer their services to anyone trying to link remote locations to a computer.

The value-added network functions of most interest to publishers are those with a heavy component of substantive information. Information retrieval systems, for example, allow users at their own terminals to display or print selected information out of a data bank. The most wide-ranging current data library, the Lockheed Dialog system, provides information from more than 150 bibliographic and factual data bases. Another data library, Lexis, allows lawyers to find and retrieve precedents and laws. Still another, Data Resources Inc. (DRI), can generate about 10 million economic time series from its data base.

These various data libraries are not necessarily reached by separate networks. The heart of the work of an electronic publisher is to put a store of information in a computer's memory and to arrange for customers to be able to consult those files. The information stored can be anything or everything that now appears in print. The computers that store the data can be accessible to a variety of networks, such as dial-up phone lines (which are expensive), leased lines (which pay for themselves only for heavy users), cable systems, or value-added networks.

None of the publicly available data archives in the United States owns its own transmission lines. But elsewhere there are cases of vertical integration, with the same organization offering both the publishing service and the underlying carrier plant. In Great Britain, France, Canada, and Germany, that is the structure of videotex services by which information bases are connected to customers' television sets by dial-up telephone lines. Prestel, the British videotex service, was initiated by the phone system itself, which owns the computers in which the data bases are stored as well as the communications net. However, Prestel's founders were sensitive to the dangers of a common carrier becoming a monopoly publisher, so the British rules allow any competitor to set up a rival videotex service using the British telephone system's transmission facilities. Also Prestel originates no information. Its library of data is entirely generated by other companies, who provide horoscopes, timetables, race results, theater reviews, news of the day, stock quotations, gardening hints, who's who, or whatever. The information providers are largely existing newspapers, magazines, and publishers. They pay a fee to place their information on the Prestel computer; in turn they earn a fee each time the data they supply is used. In other countries the separation of the electronic publisher from the information provider is not so sharp. The vertical flow of information can be cut in many different ways.

Yellow Pages

In the United States the issue of vertical integration between publishing and transmission has arisen in connection with electronic yellow pages. A bogy haunts the boardrooms of newspaper publishers: the possible electrocution of classified ads. If the 31 percent of advertising revenue that these ads bring in is lost to competitors offering want ads on videotex, newspapers will be in crisis.[11] An editorial or human interest story can be read more comfortably from a printed page than from a screen display; it can be read on the subway, or in bed, or in an easy chair. But an interactive terminal is far more convenient for classified reference material. Readers can call up precisely those listings that meet their needs, such as three-room apartments below $400 within twenty minutes of downtown. The listings can be instantly updated or purged. If a job is still listed in the electronic want ads, it is still open. Also the ads can be transactional as well as informational devices. An ad for a left-handed hot-pot glove can provide the code number to key in to place an order. There is little doubt that once electronic home terminals have become pervasive, they, rather than printed newspapers, will be the medium of choice for classified ads.

Less predictable are the form and organization these computer systems will take, though they will certainly be some mix of the possible alternatives. The hardware could be microprocessors sold in computer stores, or it could be bridging attachments from the phone to the television set, namely videotex. The systems could use typewriter keyboards, light pens, or voice input-output. The business could be controlled by print publishers turned electronic, by cablecasters, by computer companies, or by telecommunications carriers.

The battle over control is being played out between AT&T and newspaper publishers. One objective that the Senate, the House of Representatives, and the 1982 antitrust consent decree have each tried to achieve in its own way is to allow AT&T to move into the computer age in its customer services, which the 1956 consent decree prohibited by limiting the Bell System to communications common carriage business. But newspaper publishers, alarmed lest AT&T start providing electronic yellow pages and become the classified ad publisher of the future, lobbied hard to prohibit this.[12]

Electronic telephone directories make sense. Instead of chopping down trees to give each subscriber a large free book annually, the company would provide subscribers with keyboards on which to type in names, and then those persons' phone numbers would be displayed on a screen. Electronic yellow pages make sense, too. Instead of just looking up a listing of restaurants, subscribers could get a listing of the kind of restaurant they want, with information on the day's special or the current waiting line, and could even have an interactive way of making a reservation. The French telephone system is currently running an experiment with thousands of customers trying out on-line electronic directory services.

But under a clause that the American Newspaper Publishers Association (ANPA) lobbied into Senate bill S898 in 1981, electronic yellow pages are not allowed: "AT&T or any affiliate thereof . . . may not provide (within any area in which AT&T is providing exchange telecommunications service) cable service, alarm service, mass media service, or mass media product." This bill applied the restriction on publishing to those who ran the local exchanges; but the following year the Department of Justice's eight-year antitrust case against AT&T ended in a consent decree divesting the local operating companies from AT&T, thereby leaving the latter free to enter the electronic yellow-page business. The ANPA promptly persuaded the House Committee to draft a bill prohibiting any telephone carrier from publishing information over its own lines, a bill intended to bar AT&T from using its long lines for anything it publishes. Finally, the ANPA successfully appealed to Judge Harold H. Greene to amend the 1982 consent decree by adding a provision forbidding AT&T from entering into media businesses for the next seven years.

This story has several disturbing features. One is its First Amendment implications; another is its King Canute character. Supporters of the restriction on AT&T argue that it does not abridge the right to publish, since AT&T may publish anything it wishes so long as the material is not transmitted on its lines. But this argument is technologically ignorant.

The restriction implies that an electronic publisher sends out information like mail over some specific delivery channel. In fact, however, electronic publishing generally works the other way around. The publisher stores the information in a computer. The customers gain access to that computer by any circuits they choose. The inquiry traffic that flows over the national network of electronic highways leaves no tracks or traces which allow the computer receiving the inquiry to know what circuits were used. If AT&T loads a computer with information and allows access to it by local phone lines, which under the consent decree will not be its own, there is no way to ensure that a subscriber in another city has not used AT&T's lines to reach the local exchange. If there were a way to control this, it would be the freedom of the inquirer, not just that of the would-be publisher, which Judge Greene would be abridging.

To get the prohibition on electronic yellow pages adopted, the press raised the specter of AT&T becoming the nation's monopoly publisher. In the name of pluralism, the press persuaded the government to designate who may and may not publish. One might have expected the press in its own interest to advocate loudly the constitutional right of AT&T, like anyone else, to publish whatever it wishes. There is nothing new, however, in newspapers failing to see that the First Amendment applies to their competitors too. When radio was young, the press in both Great Britain and America sought to hamstring it by censorship.

If newspapers could in fact preserve classified ads for themselves, their brief infidelity to principle might be forgiven, since an important and valuable institution, the press, would be aided by preserving the status quo. But no act of Congress denying a publisher's license to AT&T will keep classified ads from going electronic. Although someone other than AT&T may become the ultimate vendor, the convenience of electronic publishing for reference material is so great that nothing will prevent the loss of such material by newspapers.

American newspaper publishers understand this fact. They nevertheless campaign for a violation of the First Amendment because a five-or-ten year delay in the emergence of a competitor means much to an investor, and this they may achieve. Also newspaper publishers in the form of media conglomerates may become the ones to provide the electronic want-ad service. Perhaps newspapers as they now exist may shrivel somewhat, but to their publishers this is a far less painful prospect if they are to be the entrepreneurs who sell the follow-on information service.

There is a plausible aspect to the newspapers' case, namely the prospect that carriers may discriminate against other publishers to favor their own publications. Some countries could well decide, in a concern for freedom, to create a system in which one limited class of companies, namely communications common carriers, would be prohibited from publishing. In countries where carriers are governmental monopolies, as they are in most of the world, the case for that decision would be strong. But in the United States the telecommunications carriers are private, and the constitutional norm of the First

Amendment prevails. The unique American tradition provides no ground for a distinction among would-be publishers. The constitutional prohibition on congressional abridgment of the right to publish is flat and general.

Yet in the United States too, if one believed the ANPA argument that allowing AT&T its First Amendment rights meant ending up *de facto* with a single monopoly publisher against whom none could compete, a rational person might say, "Violate the First Amendment in order to preserve it." But this argument is flimsy. AT&T is a common carrier. By both common and statute law, it must serve all comers, competitors included, without discrimination. Western Union gets phone service from AT&T; AT&T sends bills out by mail; the postal service has telephones in its offices. None may deny service to its competitors. There have, of course, been violations, prosecutions, and damage suits. MCI won a triple-damage judgment of $1.8 billion from AT&T for discriminations against it in service. Though that sum is being reduced on appeal, the common carrier law has proved enforceable; it works.

So even if AT&T publishes information, in electronic yellow pages or otherwise, it must in a common carrier system make its transmission lines equally available to others who have information to publish. What is required is a stern legal prescription of unconditional interconnection. The situation is no different for print. A free press depends on obliging the postal common carrier to accept everything, including publications against the postal system's owner, the government. An electronic free press depends on similarly requiring carriers to interconnect with everyone, including the carriers' competitors.

Monopoly and Competition in Network Services

Issues parallel to that of the yellow pages arise in a number of other contexts. Two examples are telephone Dial-It services and videotex services on cable. Dial-It, offered by AT&T, is a nationwide recorded message service like the local one for time and weather. For fifty cents one dials a "900" number and hears, for example, the latest sports results. This is a publishing activity whose future under the consent decree is now in question. Dubious as it should be under American law for this activity to be forbidden, it is certainly valid to require that all would-be competing publishers have full access to all

AT&T facilities which Dial-It uses. This includes not just the phone lines but also the billing system. Any competitor with a tape recorder could offer information equivalent to that on Dial-It, but when anyone phoned in, the competitor lacking use of the carrier's billing computers at the phone switch would have no way of charging.[13] To solve this problem, the FCC has required AT&T to offer such competitive services as Dial-It through a separate subsidiary which has no advantage over competitors. At the local level the same problem arises with time and weather recordings or Dial-a-Joke. If some entrepreneurs think that there are better jokes to be sold, lines and billing computers for the service should be fully available to them, but there should also be no ban on jokes by the phone company.

Exactly the same problems will arise if cable systems become the carriers for videotex. Suppose a cable operator offers a videotex service on one channel which allows viewers to tap in a number on a key pad to get recipes, or another number to get the latest news bulletins or other information services. A cable entrepreneur who had an interest in a cookbook publisher or news publisher would have a strong motivation not to lease space to a competing cookbook or news publisher on the system's videotex computer. But even if such a lease were made, the cablecaster would have a competitive advantage unless the terms of leased access included all the billing and payment facilities. Without nondiscriminatory access, other publishers will be frozen out.

In an analogous situation outside the domain of speech and press, transportation common carriers are often prohibited from going into the express business in competition with their customers, which makes it easier to enforce the nondiscrimination requirement. The argument for such a regulation on commerce is strong, but in the United States the First Amendment would seem to bar the use of this device for communications. In the case of publishing, a law that excludes anyone from the business is palpably unconstitutional. The alternative is strictly to enforce nondiscriminatory access.

There is precedent for this alternative even in the domain of print. The courts on numerous occasions have faced the problem of the relationship of the First Amendment and the antitrust laws as they apply to publishing. Antitrust laws, like tax laws and labor laws, apply to publishers as they do to any other business. In 1945 the Associated Press was held to be in violation of the antitrust laws for prohibiting service to nonmembers and allowing members to blackball competitors for membership.[14] Said Justice Black: "It would be strange indeed, however, if the grave concern for freedom of the press which prompted adoption of the First Amendment should be read as a command that the government was without power to protect that freedom. The First Amendment, far from providing an argument against application of the Sherman Act, here provides powerful reasons to the contrary. That Amendment rests on the assumption that the widest possible dissemination of information from diverse and antagonistic sources is essential to the welfare of the public . . . Freedom to publish means freedom for all and not for some." Three years later a newspaper in Lorain, Ohio, where it had been the sole advertising outlet, was held in violation of the antitrust laws when it refused ads from firms who advertised on a new radio station.[15] These cases involved monopolization of a bottleneck. The courts have held that a business that controls a scarce facility has an obligation to give competitors reasonable access when the facility is essential to the competitors' business.[16] Where monopoly exists, access is the normal legal requirement.

Monopoly, however, will not be the usual situation in electronic publishing. If access to the basic carriers exists, entry is easy to the electronic information market. Electronic publishing over common carriers may be as competitive as print publishing or even more so. Anyone with a computer and some organized information located on it can offer the information for sale. The customers are as close to the data base as to their telephone. Publishing of information is thus likely to become a more competitive industry than it is today.

Insofar as hardware affects the outcome, no natural monopoly seems likely to exist in electronic publishing if the publishers have access to basic carriers. Any number of computers in which data is placed can be reached by customers over value-added networks, and the efficiency of storage computers does not increase with size. Software complicates the picture; in that area there are natural monopolies. Once an organization has compiled a bibliography of all the chemical journal articles over the past twenty years, no other sane entrepreneur will duplicate that massive effort. If real estate agents get together to share a consolidated listing service, no one on the outside can compile so comprehensive a list. If two companies compete to offer a consumers' information service that lists prices of standard commodities in local stores, their coexistence will be unstable; if one is more reliable or more comprehensive than the other, it will draw more customers. Since the cost of compiling the data base is essentially a constant, the larger company can divide that cost over more customers, charge less, draw still more customers, and ultimately become a monopoly. A specialized information base is often a natural monopoly.

Such monopolies, however, are narrow. Competitors will come along to offer slightly different services. The chemical bibliography may have both too much and too little information for biochemical engineers, leaving room for a more specialized bibliography for them. The real estate agents may not cover rooming houses or hunting lodges; there may be room in the market for such specialized services. The consumer product service may do a good job on prices, but fail to give product test information about quality. Competitors can enter the market with differentiated publications.

A Look to the Future

The electronic information marketplace of the future will thus be a complex, changing arena of monopolistic competition among many players—probably more than in print publishing today—where narrow natural software monopolies are reinforced by copyright. Such pluralism is at least possible if public policy does not prevent it. Technology will provide a variety of delivery vehicles and forms of information display. The content that publishers will offer over these vehicles will be in voice and video, in addition to text. The messages delivered can at some cost be multimedia spectaculars, but even at low cost they can mix audio and visuals with text. Today one can buy a book or magazine, perhaps beefed up with pictures, and one can buy a cassette or disk of a movie, perhaps clarified with captions; but electronic technology will create options for mixes not yet dreamed of. Entrants into electronic publishing may therefore come from many directions—not only from book, magazine, and newspaper publishing, but also from cablecasting, television, Hollywood, the computer industry, and the telecommunications industry. As long as all these vigorous competitors can obtain access to the major carriers, monopoly among electronic publishers will not arise easily.

Although paper will not vanish, electronic publishing may in the long run evolve to something radically different from what we know today. Though pluralistic, competitive, and economical, like print, it may differ markedly in content from what is now found in magazines, newspapers, and books. Automobiles looked like horseless carriages at the start but not forever; it may be the same with electronic publishing.

One change that computers seem likely to cause is a decline of canonical texts produced in uniform copies.[17] In some ways this change will signal a return in print to the style of the manuscript, or even to the ways of oral conversation. Since Gutenberg, books, articles, manuals, and laws have been available in diverse locations in identical form. From this availability followed footnoting. When the work and the edition are named, anyone anywhere can locate an identical version. From this also followed cataloguing and the compiling of bibliographies. The identification of a printed publication by an entry in a catalogue or bibliography is unambiguous. One can even say with some precision whether a bibliography in a particular field is fairly complete or not.

Contrast this situation to the world of manuscripts where every copy was unique, with its own variations. Variations engendered a central problem in scholarship; careers were devoted to inferring the original version. Scribal rituals focused largely on preventing deviation, and in some traditions, error in transcribing a sacred manuscript was a sin.

The contrast is even greater between the modern canonical book and the world of oral dialogue. Neither in Plato's Academy nor in a modern seminar does scholarly procedure allow for cataloguing or footnoting of the ideas expressed. There is no fixed unit in which ideas occur. Their flow is amorphous.

Electronic publishing returns to these traditions. A small subculture of computer scientists who write and edit on data networks like the Arpanet foreshadow what is to come. One person types out comments at a terminal and gives colleagues on the network access to the comments. As each person copies, modifies, edits, and expands the text, it changes from day to day. With each change, the text is stored somewhere in a different version.

Computer-based textbooks may exist in as many variants as there are teachers. All teachers on occasion desire to correct or modify the textbooks they use; if texts are in a computer, they can and will do that. Each teacher will create a preferred version, which will be changed repeatedly over the years. Or in a literature or drama course one exercise might be to take a text and try to improve it. Reading thus becomes active and interactive. Penciled scribbles in the margin become part of the text and perhaps even part of a growing dialogue as others agree or disagree.

There are problems with this kind of fluid dialogue. Scientists, who were among the earliest users of computer information networks, are of two minds about the possible displacement of refereed journals by the more informal flow in computer-linked interest groups.[18] In many endeavors in law and humanities as well, one needs to identify the original or official version. Conventions will certainly develop for constraining and labeling variant versions, but they cannot stop proliferation. Before a person can read a text on a computer terminal, the text must somehow be stored in the computer's memory. Once there, it can be copied, modified, and retransmitted at will.

The implications for scholarship are mind boggling. Blue sky writings on the wonders of the computer age often describe how scholars will be able to call up instantly at their terminal any book or article from the world's literature. Wrong! The proliferation of texts in multiple forms, with no clear line between early drafts and final printed versions, will overwhelm any identification of what is "the world's literature."

For copyright, the implications are fundamental. Established notions about copyright become obsolete, rooted as they are in the technology of print. The recognition of a copyright and the practice of paying royalties emerged with the printing press. With the arrival of electronic reproduction, these practices become unworkable. Electronic publishing is analogous not so much to the print shop of the eighteenth century as to word-of-mouth communication, to which copyright was never applied.

Consider the crucial distinction in copyright law between reading and writing. To read a copyright text is no violation, only to copy it in writing. The technological basis for this distinction is reversed with a computer text. To read a text stored in electronic memory, one displays it on the screen; one writes it to read it. To transmit it to others, however, one does not write it; one only gives others a password to one's own computer memory. One must write to read, but not to write.

Or consider the case of computer-authored texts. A computer program may operate on raw numerical data and from that data generate a readable report on trends, averages, and correlations. An-

other program may operate on articles and, without human intervention, generate abstracts of them. Certainly the computer program
that does this is itself a text and is copyrightable under present law.
But what of the text that the program generates? Who is the author
of the computer-written report or abstract? The computer? The idea
that a machine is capable of intellectual labor is beyond the scope of
the copyright statutes. Can a computer infringe copyright?

In short, the process of computer communication produces multitudinous versions of texts, which are partly authored by people and
partly automatic. The receivers may be individuals, or they may be
other machines that never print the words in visible form but use the
information to produce something else again. So some of the text
that is used exists electronically but is never apparent; some is
flashed briefly on a screen; and some is printed out in hard copy.
What starts out as one text varies and changes by degrees to a new
one. Totally new concepts will have to be invented to compensate
creative work in this environment. The print-based notion of copyright simply will not work.

One case which illustrates both the kind of copyright problems
that will increasingly arise and the direction of solution to them is
still in contention. The Board of Cooperative Educational Services
(BOCES) of Erie County, New York, taped video programs of educational value off the air and sought no permission to do so. BOCES
circulated to the county schools catalogues of the tapes available and
sent the tapes out for classroom use on request. Encyclopedia Britannica's educational film company sought and got an injunction for
copyright violation. Judge John T. Curtin of New York would not
concede that BOCES, an organization with a staff of more than five
persons and a half-million dollars of equipment, all devoted to taping off the air, was engaged in fair use, even if the taping was not for
profit and was for use only within the county's schools.[19]

By the classical notion of copyright, the film company acted as it
had to in seeking to stop those who copied its material without permission. But with videocassette technology the effort is futile.
BOCES may be closed down, but nothing can stop scores of Erie
County teachers and hundreds of students from using their own re-
corders to tape off the air whatever programs are identified as worth
studying. And these copies can be copied in a system quite out of
control. However, if an organization like BOCES makes its services
valuable to the schools by following the program schedules, inform-

ing the schools on what the programs deal with, and doing the work of taping and archiving, then a natural bottleneck will exist at which the program producers can collect compensation.

Service organizations like data and tape libraries both do something from which the ultimate users get value and themselves have an organization and investment that is vulnerable enough to motivate them to develop a responsible legal relationship to the producers and publishers. The existence of a finite number of such bottleneck nodes can, like print shops of the past, make copyright work. Organized service functions for which users pay are the key to a compensation system for the era in which copying is easy but consumers need help in using the complex and overwhelming volumes of information. Britannica was perhaps right in the short run in the tactics it followed to protect its immediate legal interests but was disastrously wrong in regard to its own long-term interests. It should be seeking to foster a system of service organizations like BOCES, not destroy them.

Computer-authored texts have other radical implications besides those for copyright. Before the computer, every communications medium was essentially dumb. If it worked, it delivered at the far end exactly the message that a human being had inserted at the start. There could be noise or attenuation, or the paper could tear, but the medium added nothing positive. What a human being put in, a human being could take out, and that was all. Now for the first time the message that goes in is not necessarily the message that comes out. For the first time, thanks to automated operations using digital logic, messages may be modified or even created in the machine.

Take a very simple example, the airline reservation system. A passenger asks the clerk for a change of route. The clerk punches in a few letters to identify the passenger, the present flight plan, and the one desired. It is a short message, and the clerk gets back a short message composed by the computer confirming or rejecting the desired reservation. In the process, however, a large number of other messages may have been generated that no human being ever sees. The computer of the passenger's airline may inquire of the computer of some other airline about a connecting flight; it gets an answer back; it evaluates the reply; and it may inquire further if the desired seat is not available. Human beings write and see only a small part of the total traffic.

Today when one goes touring, one buys a guidebook not too bulky to carry and skips around in it to read about the places where one goes. A computer-controlled guide disk could enable travelers to reassemble text, pictures, and maps from a shelf of information in the order of their trip and according to their taste for historical, sociological, or culinary details. Today when people buy an art book, they look at pictures someone else has chosen. Or if they buy a Vasarely book, they have sheets of plastic to superimpose for changing effects. But a computer-published art program of the future might be a kind of synthesizer allowing the lay reader to explore colors and perspectives that change as the reader mans the controls. The products that publishers sell may increasingly change from frozen texts to interactive tools.

Electronics is changing publishing only gradually. A reporter today writes a story on a word processor, whereby editing and page makeup are computer aided; but what comes out at the end is a newspaper looking just the same as it always did. Eventually electronic publishing may become more like an electronic game, permeated by lights and sound along with words. The players will initiate, and the machine answer back, in an interactive conversational process. It may be for fun, for management of daily life, or for work. Whatever it is, it will probably in the end resemble present-day publishing no more than does the business or product of today's Time-Life conglomerate resemble that of the scriptorium of a monastery.

Implications for a Free Press

If such drastic changes occur in the publishing industry, great changes could occur in First Amendment practices too. The industry is a complex structure imposed on a flow that starts in institutions of culture and science where knowledge and art are produced, and goes through institutions where editing and packaging take place, to printing plants and then distributors. In the print tradition of the First Amendment the entire flow has been free of government controls. In various cases about academic freedom and reporters' rights, the search for knowledge has been recognized by the courts to be part of speech. So too has the integrity of records and files. The courts protect the acts of speaking, parading, picketing, distributing handbills, and selling publications. They also protect the physical plant which is used; printing presses are beyond the reach of government licensing.[20] For print communication the one element of the flow that was a monopoly was the post office, whose authority the courts have curtailed so as to protect the freedom of the users.

Among print publishers the structure of the flow ranges all the way from almost complete vertical integration to almost complete disintegration. Newspapers are typically at the integrated extreme. Stories are researched and written by their own hired staff, they have their own printing plants, and they often bypass the postal monopoly, delivering papers through their own trucks and newsboys. At the other end are publishers who have only an office and no plant at all. They contract with authors, printers, binders, and distributors, and themselves perform just an organizing function.

Electronic publishing has to carry out all the same functions and may do so in an equal variety of ways. In electronic publishing, as in print publishing, only one—and the same one—of these various linked functions is generally monopolized, the physical transmission. Electronic publishers depend on access to the basic carrier just as print publishers depend on access to the postal system.

Whether that monopoly element will be more important or less in the electronic system than it is in the print system is likely to vary with time. Postal delivery was more important for newspapers in the early years of the country than it is today. The telephone and telegraph monopolies were much stronger before there were alternative technologies for transmission, such as microwaves, satellites, and cable. Monopoly seems likely to be even further eroded in the future.

Despite these technical forces in favor of competition and the FCC's recent policy favoring competition even in physical carriage, the degree of monopoly is likely to remain significant in transmission plants.[21] There will continue to be only one microwave link on many routes and almost everywhere only one cable television trunk and one set of telephone wires. Common carrier principles, including compulsory nondiscriminatory access, will continue to be appropriate to these parts of the transmission plant, but by analogy with print publishing, it seems plausible to expect that all the other elements of electronic publishing will, by constitutional requirement, be fully unregulated. Just as with print, electronic publishers might set themselves up at will with various patterns of vertical integration. Some electronic publishing organizations would, like newspapers, carry on most activities in-house. Other publishers would, on the analogy of book publishing, be vertically disintegrated. The British Prestel videotex system is more like the former; the United States electronic information market is more like the latter, with data base preparation, archiving, computer networking, and physical transmission all now usually done by different companies. It might be assumed that, except where there are elements of monopoly, the government would have no say about how electronic publishing is organized and done; that would be left, under the First Amendment, to be determined by the free marketplace of ideas. But this is certainly not the case today.

How Networks Are Now Regulated

Neither in the United States nor in Europe has electronic publishing been left unregulated. Even those parts of the electronic publishing system that simply make use of subscriber channels on the underlying physical plant and which can be totally competitive in character have been extensively regulated. The position of the FCC under the 1934 Communications Act and of almost every foreign telecommunications authority (usually called PTT for post, telephone, and telegraph) is that, insofar as a value-added network is a communications carrier, it may not operate without permission.

In Europe the usual position is that if a value-added network (VAN) is in the business of switching messages between two other parties, it is violating the post office monopoly, and this is not allowed. For example, the public packet networks Telenet and Tymnet wanted to install switching in Europe so that a customer needing to use a data base in the United States could do so by dialing up that European switch using the PTT's lines, after which the query would travel on a high-speed line, leased jointly from the PTT of the country of location of the switch and the American international record carrier, to the Telenet or Tymnet node in New York and from there over that network's lines, leased from AT&T, to the data base; the reply would follow the reverse course. This, however, was not allowed. Telenet and Tymnet may not put switching nodes of their own in Europe; switching has to be done by the PTT. So the arrangement is that in each European country the PTT installs the node under its own legal control, though in some instances Telenet or Tymnet actually puts it in as a contractor; the message legally enters the VAN's packet network only in New York, traveling there under PTT and international record carrier auspices. The main difference as far as users are concerned is that they pay more, since the PTT sets rates for use of its switch and its service well above those that the packet network, acting as wholesaler, would have charged. The PTT thus has protected both its revenue and its monopoly status.

One might expect American authorities to adopt quite different policies regarding monopolistic underlying physical networks, on the one hand, and generally competitive value-added networks, on the other. The FCC is moving in this direction, but from the 1934 Communications Act until well into the 1970s this distinction between the two types of carriers was not fully recognized. One might also expect American authorities to have taken the view that the owner of monopolistic physical transmission systems should be a common carrier. The authorities have not done so consistently. This policy will continue to be applied to local phone companies under the 1982 AT&T consent decree. It was applied in the past to the entire telephone network and will presumably continue to be applied also to the long-distance portion. However, cable networks have not been set up as common carriers.

Since value-added networks are not normally monopolies, it might be expected that they would not be regulated in the United States. However, under the Communications Act of 1934, electrical communications networks must have authorization from the FCC. In 1934 the concept of a VAN did not exist, and it might be argued that what Congress had in mind was underlying physical networks, like the telegraph and telephone networks. However, the FCC interpreted the law as requiring it to decide whether *any* communications carrier offering public service is in the public interest, and courts sustained it in that. In 1976 the commission ruled that value-added carriers must be regulated like the basic carriers.[22] Then in 1980 the FCC tried to reverse itself. While continuing to claim that it was required to exercise jurisdiction under the 1934 act, it decided that it would regulate VANs by forbearing from regulating them. Thus in principle, the FCC's claim that it is the judge of whether there are already enough carriers in the business or whether a new one is a good idea is so far uncompromised. For VANs that are essentially publishers, it is hard to sustain the constitutionality of this position.

Such authorization authority is claimed by the FCC only if the service that the VAN provides is communication. In the 1960s and 1970s remote time-sharing became an important part of the competitive unregulated computing industry. Computer networks using the phone lines were established. These had the incidental ability to move messages in competition with posts and telegraphs. This capability created a regulatory issue. Should time-sharing systems be banned, as they are in many countries, from providing a useful new message service? Or should they be allowed to operate in competition with common carriers without bearing any of the obligations and regulations on common carriers? Or should they be forced into a generally undesired regulatory mode, even though they have none of the monopoly characteristics that justify common carrier regulation, simply because they compete with a previously regulated industry? In its Computer Inquiry II the FCC found no way to distinguish computing on a time-sharing network from communication over it. They therefore had to achieve their purpose of not regulating computing by exempting it as a new category of enhanced communication service.

This decision evaded what before had been a disturbing First Amendment problem. As long as time sharing and communicating were technically indistinguishable, and as long as using a computer network to communicate without a license was illegal, it would have required intrusive controls to ensure that citizens using their terminals not communicate over a computing facility which had not been licensed as a carrier. Airline reservation systems all over the world, for example, are prohibited from carrying messages not related to airline traffic. For a reservation clerk to send a message to waiting relatives that Mrs. Smith has missed her flight, or for one clerk to send "happy birthday" to another, is in most countries illegal. So also is sending messages among banks on a funds transfer system. Or consider the way I wrote this book at a terminal. Editing, using the computer to modify the text, was undoubtedly within the law, but sending a chapter from one point on the network to a colleague to read at another may have fallen afoul of it. It was using the time-sharing network as a means of publication, transmitting views to others on the system, and so it was treating the system as a communications carrier. The computer was in that use a printing press, which it was not supposed to be. It is hard to conceive of a legal conception more blatantly unconstitutional.

The FCC in Computer Inquiry II changed those rules by deregulating enhanced communications, which includes all these computer-based applications, but it did so not on constitutional grounds. It deregulated most communications carriers on the grounds that they are operating in a market that can adequately protect the consumer. In so doing, the FCC explicitly claimed the authority to regulate but chose to forbear from doing so.[23] The FCC found that tariff approval, market entry and exit controls, and similar regulations are needed only for "basic" communications services provided by companies that are dominant in their portion of the market. Basic communications service refers to the transmission of a message from one point to another without changing it, in contrast to "enhanced" service in which the carrier or VAN manipulates the message in some additional way for the customer. In the basic domain, when there is a dominant company or monopoly, the FCC continues tariff and entry regulation. For the rest, the FCC now forbears from such regulation, allowing the market to control.

Such deregulation may work well in the short run, but it does not come to grips with the constitutional issue. The lightened regulations remove burdensome requirements from VANs and other nondominant carriers. The FCC no longer requires would-be electronic publishers to prove that the public interest, necessity, or convenience requires the network service that they plan to provide; the marketplace will determine if the public wants it. But the only thing that leads the FCC to refrain from control is its benevolent judgment. The precedents for prior restraint, granting of permission, and exercise of control all stand. The FCC still keeps its eye on all carriers and telecommunications services, basic or enhanced. It still asserts its authority to step in and regulate if things go wrong. It does not recognize that, where content is involved, network usage is publishing and constitutionally may not be controlled; the 1934 act or any other communications act must be so interpreted.

That disregard of First Amendment principles in the muddle over computing and communications was not aberrant is illustrated by parallels in the story of the licensing of satellite earth stations. The Communications Satellite Act of 1962 requires FCC licensing not only for satellites and satellite radio frequencies but also for earth stations.[24] For those earth stations that are senders on the uplink, the reason is obvious. They are radio transmitters and, as such, have to be licensed to use specified frequencies so as to avoid interference. But for those earth stations that are simply passive receivers, no such justification of licensing exists.

A parallel issue arose in the early days of radio. Many countries require a radio set owner to obtain a license. The license fee is often the way in which broadcasting is financed. In the United States, however, the government never felt itself justified in licensing receivers. Divulging of radio or wire messages not intended for one was forbidden, but in accordance with the First Amendment tradition, no attempt was made to impose prior restraint by requiring a person to have a receiving license as a way of controlling eavesdropping.[25]

For receive-only earth stations, however, licenses were required. The justification was generally made on narrow technical grounds rather than on grounds of monumental dangers of the kind that might permit prior restraint. Licensing, it was said, protected buyers of earth stations from acquiring inferior equipment. Also, since a license is a kind of contract between the earth station owner and the FCC, the FCC does not license interfering transmissions at the frequency that the receiver is licensed for; the licensed earth station owner is protected. But these grounds hardly justify making the use of an unlicensed receiving station a crime. Some earth station buyers may indeed want the protection of a license, but those people who operate an earth station at their own risk without such protection are within their rights.

By 1979 the FCC came to recognize the logic of this argument, and in the prevailing mood of deregulation it eliminated the requirement of licenses for receive-only earth stations.[26] Licensing became voluntary, available to those who wanted its protection from interference. But once again no recognition was given to the fundamental consideration that under the Constitution compulsory licensing of a communications facility, except in exceptional circumstances—such as radio transmission, where the communicator may do demonstrable harm to others—is outside the authority of government. The FCC conceded not one iota of its authority in concluding that compulsory licensing of earth stations is unnecessary, not unconstitutional.

The Roots of Myopia

The roots of such an unconstitutional assumption of regulatory authority lie partly in perceiving publishing via electronic networks through the tradition of telephony and, before that, of the first common carrier, the post office. Regulation of VANs came about by unthinking extension to them of the rules governing monopoly telecommunications carriers. The special circumstances that justified exceptions to First Amendment practices for the postal, telegraph, and telephone monopolies have been long forgotten. A series of precedents established in those circumstances have been carried over to a normal domain of competitive speech where no such exception is needed.

The unconstitutional practice of licensing value-added communications carriers also developed because they are engaged in a kind of communication that was perceived as business rather than speech. The Communications Act of 1934 and its predecessors were based on the commerce clause of the Constitution. The First Amendment, however, blocked out from the jurisdiction of Congress one significant area of interstate commerce, namely human intercourse in ideas. Traffic in newspapers, for example, is not subject to regulation under the guise of commerce, though commerce it certainly is. Data traffic is a form of press, every bit as much as is the product of a printing press, yet it has not been perceived that way. It was in the decade of the 1960s, when computers and computer networks were first coming into common civil use primarily in business applications, with devices mostly manufactured by a company called International Business Machines, that the Supreme Court toyed with the notion that business communication was not protected speech under the First Amendment. This notion was finally thrown out by the Court in 1976. Nonetheless, in a period of uncertainty the notion of commercial speech took its toll in how the courts and regulators thought about computing. Just as the early telegraphs were misperceived as essentially business machines rather than presses, the computer too was easily misperceived as just a numerical calculating machine or at best a business machine.

Today, large information services flow increasingly over computer-switched networks. So do the press wire services. Publishers compose magazines and books on the same facilities. Scholars use computer networks for teleconferencing. So do people engaged in discussing public affairs. Citizen complaint and inquiry services in some places use computer networks as carriers. The membership lists and files of labor unions, political groups, universities, research institutes, and religious organizations are frequently maintained on some service bureau's remote computer. With the rapid introduction of communicating word-processing typewriters to the offices of such organizations, the computer networks to which they are attached will become their means for substantive communication. Information retrieval systems are becoming major libraries, and terminals are moving into citizens' homes. Soon most published information will be disseminated electronically.

Networked computers will be the printing presses of the twenty-first century. If they are not free of public control, the continued application of constitutional immunities to nonelectronic mechanical presses, lecture halls, and man-carried sheets of paper may become no more than a quaint archaism, a sort of Hyde Park Corner where a few eccentrics can gather while the major policy debates take place elsewhere.

9. Policies for Freedom

As computers become the printing presses of the twenty-first century, ink marks on paper will continue to be read, and broadcasts to be watched, but other new major media will evolve from what are now but the toys of computer hackers.[1] Videodisks, integrated memories, and data bases will serve functions that books and libraries now serve, while information retrieval systems will serve for what magazines and newspapers do now. Networks of satellites, optical fibers, and radio waves will serve the functions of the present-day postal system. Speech will not be free if these are not also free.

The danger is not of an electronic nightmare, but of human error. It is not computers but policy that threatens freedom. The censorship that followed the printing press was not entailed in Gutenberg's process; it was a reaction to it. The regulation of electronic communication is likewise not entailed in its technology but is a reaction to it. Computers, telephones, radio, and satellites are technologies of freedom, as much as was the printing press.

Trends in Communications Technology

The technologies used for self-expression, human intercourse, and recording of knowledge are in unprecedented flux. A panoply of electronic devices puts at everyone's hand capacities far beyond anything that the printing press could offer. Machines that think, that bring great libraries into anybody's study, that allow discourse among persons a half-world apart, are expanders of human culture. They allow people to do anything that could be done with the communications tools of the past, and many more things too.

The first trend to note is that the networks that serve the public are becoming digital and broadband. Today, the only broadband signal received by the ordinary household is its television picture. It is sometimes questioned whether there are any other uses for which an end-to-end broadband digital network available to every household and workplace will be demanded. Such a network would allow two-way transmission of high-definition pictures and text in whole volumes at a time, along with voice, videotex, and other low-speed services, but why should that be wanted? There are, indeed, good reasons. High-definition pictures are not just fun. As manual mail service gets less reliable and more expensive, the sending of magazines, catalogues, videotapes, and videodisks electronically rather than physically will become an attractive option. Nor is text delivered in whole volumes at a time just a luxury. Text a page or so at a time, even if it comes faster than one can read, is satisfactory for electronic mail or for retrieving pages one knows one wants, but it does not do for browsing. To use a terminal the way one uses a bookshelf or filing cabinet, one must be able to thumb randomly through thousands of pages. And when computers talk to computers, even though the size of the files they flip back and forth may be modest, a second is too long for them to wait; their bursty traffic requires large bandwidth for short periods. Millions of offices and homes may have computers and want bandwidth enough for them. So if people at home or work want high-definition moving pictures, if they desire two-way video for teleconferencing or teleshopping, if they wish to browse in libraries rather than just reading predefined pages, and if they compute, then the demand for end-to-end broadband networks will exist.

To serve the public, there will be networks on networks on networks. Separate nations will have separate networks, as they do now, but these will interconnect. Within nations, the satellite carriers, microwave carriers, and local carriers may be—and in the United States almost certainly will be—in the hands of separate organizations, but again they will interconnect. So even the basic physical network will be a network of networks. And on top of these physical networks will be a pyramid of service networks. Through them will be published or delivered to the public a variety of things: movies, money, education, news, meetings, scientific data, manuscripts, petitions, and editorials.

Another trend to note is toward increasing sophistication of the equipment on the user's own premises. Since the output and input of networks may be either printed on paper, shown on a screen, or declaimed in sound, the equipment needed on the customers' premises will be costly. Although the costs of computer logic, memory, and

long-distance communication are falling, the uses that people want
to make of them are expanding even faster. A $4000 microcomputer
can today do things that would have required a million-dollar computer a few years ago, when few would have predicted that millions
of ordinary people would spend $4000 for that home gadget. In the
future, many millions of households will similarly desire large-size
high-definition screens, cameras to originate video, and large memory devices to retain libraries of information for work and pleasure.

American industry is speculating that the percentage of dispos-
able income spent on information activities will grow. Companies
are positioning themselves to be in that industry. Banks like American Express and manufacturers like Westinghouse are investing in
cablecasting; companies like Boeing are selling time-sharing services; and storekeepers like Sears Roebuck are experimenting with
videodisk catalogues. Investors see the biggest dollar growth not in
transmission or its hardware but in software and the equipment located on the customer's premises. This conclusion is what led AT&T
to accept divestiture of its local phone companies in exchange for the
freedom to sell information services and equipment to final customers. The science fiction version of the information work station of
2001 with beeps and sirens, flashing lights and video screens, may be
fantasy, but the point is right: that is where expense will lie.

Paradoxically, big customers and decentralization will both gain
from the development of more elaborate terminal equipment. However splendid may be the homeowner's equipment, it will be only a
humble version of what will exist in plants and offices. Companies
with information service and carrier billings in the millions will invest in their own networks, leased circuits, compression devices, and
other marvelous gadgets designed to help them operate efficiently
or cut costs. Depending on the structure of the vendors' and carriers'
tariffs, different alternatives will pay off. One trade-off will be between buying communications capacity so as to improve management control and buying local processing power so as to cut communication costs. Trends between such centralizing and decentralizing
alternatives may zigzag as technological and tariff changes affect relative prices, but the costs of computing equipment used to store data
locally, to compress it, and to process it will probably fall farther and
faster than the costs of transmission.

This trend favors decentralization. More and more will be done at
the distributed nodes of networks to economize on transmission.
That dispersion will be pushed farthest by big users, for they have

the resources and technical capability to do so. When in large enterprises the competence and autonomy of scattered nodes are thus strengthened and their subservience to a center is thus lessened, the result, paradoxically, may be decentralization.

Another obvious trend is that with the new technologies, the world is shrinking. To talk or send messages across the world is coming to cost little more than communicating in one's own region. The charge for a call from New York to Los Angeles is now little more than for a call from New York to Albany. Both involve identical costs for the local loops and switches, for setting up the call, and for billing. The variable cost of extra microwave links is a minor item. With satellites, distance becomes almost totally unimportant. Patterns of human interaction will, as a result, change. There will be less cost constraint to do business, consult, debate, and socialize within one's own region only. There will be more freedom to do so with anyone anywhere with whom one finds affinity.

This development, along with the development of multiple technologies of communication and of cheap microprocessors, will foster a trend toward pluralistic and competitive communication systems. With hundred-channel cable systems, videocassettes, videodisks, ISDNs, and network links to thousands of on-line information services, there should be a diversity of voices far beyond anything known today. Telephone monopolies are being broken up. Before computers, phone administrations forbade connecting any "foreign attachment" to their network; today in the United States, Japan, Great Britain, and elsewhere, customers are being allowed to buy terminals at will and to attach them. Before microwave and satellite transmission, phone administrations had a monopoly in stringing wires from city to city, but these new nonwire transmission media are often managed by different enterprises. In the United States such competition already prevails in long-distance service, and local exchange service will not long remain completely monopolistic. Digital termination service, cellular radio, and cables carrying voice and data will all compete with the local phone company.

There is no reason to assume that the communications network of the future will be a single large organization with a central brain. It may be so, but it need not be. Having a hierarchical structure governed by a central brain is only one way to organize complex systems. A human being is organized that way; so is a nation-state. But the capitalist economy is not, nor is the complex system of scientific knowledge, nor is the ecological system of the biosphere. For an uncentralized system to function, there must be some established ways of interconnecting the parts other than by command; the interconnections may be managed by conventions, habits, or Darwinian processes. Capitalist property rights are enforced by laws; language is enforced by custom; creatures in the biosphere do not survive if they cannot metabolize other species.

An uncentralized set of communications systems can function as a single system only if traffic on each network can move through interfaces onto the other networks. The critical requirements are three: the right to interconnect, conformity to technical standards that make interfacing possible, and a directory system.

The variety and autonomy of networks for special groups and services may grow rather than decline, though most of them will interconnect with each other. Some of these networks will and others will not have their own central brains. The different kinds of communication—video, voice, and text; informational and emotive; public and personal—are likely to require differently designed networks, even if interconnected.

Digital technology promotes the trend toward distributed processing throughout the system and against a central brain. It is easier to convert one system of 0,1 pulses into another such system than it is to interface the analog memoryless communications systems of the past. A directory search in the absence of a single universal list is more likely to succeed if it uses intelligent digital devices that scan associative data structures at nanosecond speeds and that communicate with all nodes at the speed of light than if it is bound by the slower circuit switching of the past.

Perhaps the most remarkable trend to note is one whereby the artificial intelligence of computers will increasingly create and read many of the messages on the networks of the future. These computer-composed messages sent from computer to computer may mostly never be seen by a person at all. In an electronic funds transfer, only a few bits are needed to say debit an account by $27.50. Most of the traffic involves checking and rechecking to see whether the signature is authentic, whether money is available, and what balance is left.

The future of communications will be radically different from its past because of such artificial intelligence. If media become "demassified" to serve individual wants, it will not be by throwing upon lazy readers the arduous task of searching vast information bases, but by programing computers heuristically to give particular readers more of what they chose last time. Computer-aided instructional programs similarly assess students' past performance before providing the instruction they need. The lines between publication and conversation vanish in this sort of system. Socrates' concern that writing would warp the flow of intelligence can at last be set to rest. Writing can become dialogue.

Such are some technical features of the communications system that is emerging. Technology will not be to blame if Americans fail to encompass this system within the political tradition of free speech. On the contrary, electronic technology is conducive to freedom. The degree of diversity and plenitude of access that mature electronic technology allows far exceeds what is enjoyed today. Computerized information networks of the twenty-first century need not be any less free for all to use without let or hindrance than was the printing press. Only political errors might make them so.

Communications Policy

In most countries the constitution sets the framework for communications policy.[2] America's basic communications policies are found in three clauses. Article 1, Section 8, gives Congress the power to establish post offices and post roads. The next clause gives Congress the power "To promote the Progress of Science and useful Arts, by securing for limited Times to Authors and Inventors the exclusive Right to their respective Writings and Discoveries." And the very first amendment prohibits Congress from passing any law abridging freedom of speech or of the press. This package of provisions provided publishers with the support they needed but barred the government from interfering with their free expression.

In the comparatively simple American society of the eighteenth century, when the media depended largely on the slowly changing technology of the printing press and when government consisted of the relatively spare mechanisms of the courts, Congress, and a tiny executive branch, communications policy issues were few. They arose most often from the ability of the government to use its fiscal powers both for and against the press. The American people did not oppose the government's use of its fiscal powers to support the press. The authorities did so through the postal system, official advertising, and sinecure appointments. The idea that government should stand at arm's length from the press developed later; the earliest federal policy was to foster the media. The other possibility that government might employ its coercive powers against the press was prohibited by the First Amendment.

As Congress, the executive branch, and the courts dealt with innovations in communications technologies and, during the two centuries following adoption of the First Amendment, sought to formulate policies appropriate to them, the Amendment's original principles were severely compromised. The three main decades of such change occurred at intervals a half-century apart. In the 1870s Congress and the courts extensively restructured postal policies, imposing censorious restrictions. Also in that decade, and shortly before it, the system of common carrier regulation of telegraphy evolved. Fifty years later, in the 1920s, radio broadcasting began. For that medium Congress required that broadcasters be chosen and licensed by the state. Then half a century later, in the 1970s, computer networks, satellites, and cable systems came into extensive use. Some of the regulatory responses to them seem quite unconstitutional.

Both the 1870s and the 1920s were decades of ambivalence about civil liberties. In the 1870s a rising reform movement about both morals and economics challenged the prevailing philosophy of laissez faire. Movements for temperance, prudery, voter registration, and labor protection clashed with ideas of minimal governance. Reformers pressed for acceptance of the regulation of mail as an instrument of censorship. The 1920s saw the Palmer raids, on the one hand, and the Brandeis-Holmes dissents and decisions, on the other.[3] The sensitivity of the Supreme Court to the First Amendment, starting in the 1920s and particularly after World War II, led it to blow the whistle and stem the trend toward postal censorship.

It was in the 1920s, however, that communications policy in the United States most seriously lost its way. Without adequate thought, a structure was introduced for radio which had neither the libertarian features of the common carrier system nor those of the free market. The assumption of the new system was that spectrum was extremely limited and had to be allotted to chosen users. In Europe the chosen user was generally the government itself; in America it was private licensees. Since only a few would be privileged to broadcast, government felt it must influence the character of what they broadcast. The broadcasting organizations, unlike common carriers, selected and produced programs, but unlike print publishers, who also select what appears, there was no free entry for challengers. So government stepped in to regulate the radio forum and shape the broadcasters' choices.

By this process of evolution there came to be three main communications structures in the United States: the print model, free of regulation in general; the common carrier model, with government assuring nondiscriminatory access for all; and the broadcasting model, with government-licensed private owners as the publishers. The choice between them is likely to be a key policy issue in the coming decades. A convergence of modes is upsetting what was for a while a neatly trifurcated system. Each of the three models was used in its particular industries and for different types of communications. As long as this was the case, the practices in some industries might be less true to the First Amendment than the practices in others, but it did not much affect those media that remained in the domain of the First Amendment. What happened in one industry did not matter greatly for what happened in another.

If this situation were a stable one, there would not be much cause for worry, for if the nation retained a free printed press through which all viewpoints could be expressed, it would not lose its freedom even if broadcasting were totally government controlled. Having print as an island of freedom might be assurance enough against total conformity to authority. But the situation is not stable.

Very rapidly all the media are becoming electronic. Previously the print media were affected, but not themselves transformed, by the electronic media. The electronic media grew and enlarged their field of action but left the older media fundamentally what they had been before. This is no longer the case. With electronic publishing, the practices of the electronic media become practices of the print media too. No longer can one view electronic communications as a circumscribed special case whose monopolistic and regulated elements do not matter very much, because the larger realm of freedom of expression still encompasses all of print. Telecommunications policy is becoming communications policy as all communications come to use electronic forms of transmission.

Soon the courts will have to decide, for vast areas that have so far been quite free of regulation, which of the three traditions of communications practice they will apply. The facts that will face the courts will be a universally interconnected electronic communication system based on a variety of linkable electronic carriers, using radio, cable, microwave, optical fiber, and satellites, and delivering to every home and office a vast variety of different kinds of mail, print, sound, and video, through an electronic network of networks. The question is whether that system will be governed as are the regulated electronic media now, or whether there is some way of retaining the free press tradition of the First Amendment in that brave new world.

Resource Constraints

Historically, some media operate by different rules under the First Amendment from those applied to publications in print because of the existence of scarcities in the resources used in producing them. Abundance and scarcity of resources are the two ends of a continuum. At one end, communication is entirely unconstrained by resources; in the middle are situations in which there is constraint, but everyone can nonetheless have some ration of the means to communicate; and at the other end the constraints are such that only a privileged few can own those means.

Conversation illustrates the optimal situation in which communication is totally without resource constraints; the only limit is one's desire. There is also sidewalk enough in most places that anyone can picket a building without excluding others from passing by, though not when hundreds want to picket in the same place at once. In practice there is no resource bar to forming a congregation to worship together, as witnessed by storefront churches and congregations that meet in members' homes. Similarly anyone can send a petition or write a letter of protest. The property required to carry out these acts is trivial.

Even in such domains, where normally anyone can communicate at will without noticeably reducing the opportunities for others, there are exceptional situations in which one person's wants do constrain others. Conversation may be abundant, but conversation with a particular partner is an imposition on that person. Assemblage as a congregation can be almost costless if in a member's home, but building a large church on a desirable lot is not. In each of these situations the cliché formula is that people have the right to do as they wish, so long as they do not interfere with the rights of others. But communication in such situations involves so few resource constraints that no special institutions are set up to deal with them.

The situation is more complex when the resources for communication, while not unlimited, are available enough that by reasonable sacrifice and effort a person can get hold of some. In this situation allocation by the institutions of property and the market becomes a useful norm. An example is the printed magazine. Even the poor, by scrimping and cooperating, can produce periodicals, and some of them do. There are church bulletins for modest congregations, labor and protest papers, adolescent club and school papers. There are thousands of little magazines with stories and poems by unknown amateurs. To convert a publication into a success requires talent, capital, and energy; if the talent and energy are there, the capital may even be borrowed.

In such situations of moderate scarcity, however, not all people can have whatever means of communication they want. The means are rationed. The system of rationing may or may not be equitable or just.[4] There are an infinity of ways to partition a scarce resource. The method may be strictly egalitarian, as that which requires all legal candidates to be offered the same amount of air time under the same terms during an election campaign. The method may be merito-cratic, as that which gives free education to those who score high in exams. The method may recognize privilege, as that which allows a descendant to inherit a communications medium or a seat in the House of Lords. The method may recognize cultural values, as that which occurs when a foundation makes grants to museums or sym-phonies. The method may reward skill and motivation, as that which allows communications institutions to earn profits that depend on their efficiency.

Each of these criteria for allocation has its value, and actual public policy represents different mixes of them. Equality may have both rhetorical appeal and a great deal of merit. Yet few people would opt for totally equal access to scarce means of communication, quite in-dependent of considerations of talent, motivation, or social value in their use.

Property rights in the means of communication are a major method of allocation, but different property schemes produce differ-ent allocations. In some property schemes, if people who have radio frequencies do not use them, their right lapses and the frequencies revert to the allocating agency. This is like the small print which reads, "This ticket is nontransferable." In a market scheme, how-ever, the owners may, within the limits of the law, pass on their re-source to someone else as a gift or as part of a deal. A market scheme is predicated on a lack of faith in administered wisdom; it treats whatever allocations exist as a starting point only. It assumes that the distributed wisdom among the property holders is greater than that of a central planner.

The law creating a market defines the mix of deals that may be made and specifies some as illegal. A person moving may sell a house, but perhaps under zoning laws it may not be turned into a tavern. A shipowner with a radio license may sell a ship, including the radio facilities, but may not turn the ship's frequency to broadcasting. A cablecaster may sell a system to someone else but, under American rules, may not sell it to the owner of a television station in the same town.

Property is, in summary, simply a recognized partition of a resource that is somewhat scarce. A market is a device for distributing the use of that property. It measures the value people attach to different uses; it allows for shifts in the uses; and it depoliticizes decisions by decentralizing them. But a market is not a single device; it is a class of devices, and public policy defines the market structure.

Where some resource is either very scarce or not easily divisible, ordinary markets function badly. Spectrum, given the way it is now administered, is scarce enough that every small group cannot have its own television station. A telephone system is indivisible in that what is needed is a universal system. In such cases of monopoly or partial monopoly of the means of communication there are problems for free societies. It was for such situations that the common carrier approach was developed in the nineteenth and early twentieth centuries. Common carriers are obligated to offer their resources to all of the public equally. In the American constitutional system this is an exceptional fallback solution. The basic American tradition of the First Amendment is either the free-for-all of free speech or the competitive market of the early press.

Since scarcity and indivisibility of resources compel a departure from the print model, it is important to estimate what major scarcities and indivisibilities will appear in the evolving electronic media. Despite the profusion of means of communication that are coming out of the technology of the late twentieth century, a number of truly scarce or indivisible elements will remain in the communications system. Despite the cliché of broadcast regulation that frequencies are exceptionally scarce, spectrum is not one of them. If spectrum were allotted by sale in a market, the prices would not be prohibitive, for there are now numerous alternatives, such as data compression or transmission by coaxial cables or optical fibers, which would become economic in the presence of relatively low costs for air rights. Spectrum is only of medium scarcity.

The orbit for satellites is today what spectrum was in 1927, something that at first glance seems inherently and physically scarce. If the technology of orbit usage remained that of the 1970s, orbital slots in the Western hemisphere would have run out by about the time this book got published, and in many other places shortly thereafter. However, techniques for orbit and spectrum saving are multiplying. The real problem is spectrum, not real estate on the orbit. There is abundant space for the satellites themselves. The difficulty is to find frequencies for communicating to and from the satellites without causing radio interference. Polarization, spot beams, time-division multiplexing, on-board switching, and direct satellite-to-satellite microwave or laser links are among the techniques that help. These require agreement to use technically efficient methods, which may not be the cheapest ones. The orbit problem turns out to be a special case of the familiar spectrum problem. To keep prices down requires agreement on and compliance with efficient standards and protocols. But at a price, as much of the resource as is wanted is likely to be available.

Though neither spectrum nor orbital slots are too scarce to be handled by normal market mechanisms, there are other more severe elements of monopoly in the system. One is the need that basic communications networks be universal in reach. If anyone is to be able to send a message or talk to anyone else, there must be universal connectivity, directory information, agreed standards, and a legal right to interconnect.

Another element of communications systems that makes for central control is the need to traverse or utilize the public's property. The social costs of not granting the right of eminent domain for transmission routes are very high. Also streets have to be dug to lay cables. These requirements affect many people who are not direct participants in the arrangements.

Finally, there are areas of natural monopoly where the larger the firm, the more efficient its operation, so in the end the smaller competitors are driven out of business. This has been the case for American newspapers. They depend heavily on advertising by merchants. Where there is more than one paper in a town, merchants find it more efficient to advertise in the larger one, and the smaller ones wilt.[5] The situation was similar when there was more than one phone company in a city; customers joined the larger system because there were more people whom they could then call. Customers would also pick the larger cable system if more than one served a neighborhood. The larger system that shared the fixed plant costs among more subscribers could charge less, and with more revenue it could offer more or better programs.

In communications, economies of scale are found especially in wire or cable transmission plant. The large investment in conduits reaching everywhere dominates the equation. There is no such strong economy of scale either in over-the-air transmission or ordinarily in programing or enhanced services. Where economies of scale and therefore natural monopolies do exist, some form of common carriage access is appropriate. It exists in phone service. Common carriage in some form may well come for cable as well.

Although there are elements of natural monopoly in both newspaper and electronic carrier markets, common carrier procedures have been applied in one and not the other. A newspaper may be the only one in its town, yet still it enjoys all the privileges of the ordinarily competitive printed media. Under the First Amendment, as interpreted in the Tornillo case, it cannot be forced to yield access to anyone.[6] The issue of whether, as a monopoly, it should be so compelled is not a trivial one. Barron's argument in Tornillo was not dismissible lightly, but the Court did reject it and continued to give newspaper owners full autonomy of editorial decision.

The fact that the newspapers have maintained such freedom from a requirement that normally goes with monopoly distinguishes them from cablecasters, who are ordinarily required by their franchises to provide some access. Historical complexities, not simple logic, account for this paradox. In both cases, but especially for newspapers, the scope of the monopoly is incomplete. At least as important is the fact that newspapers, reared in the tradition of a free press, have behaved so as to discourage the issue from arising. Newspapers, as they moved into the status of monopolies, had the wisdom to defuse hostility by acting in many respects like a common carrier. Aware of their vulnerability, they voluntarily created something of an access system for themselves. Unlike their nineteenth century ancestors, they see themselves as providing a forum for the whole community. They not only run columnists of opposite tendency and open their local news pages willingly to community groups, but also encourage letters to the editor. Most important of all, they accept advertising for pay from anyone. Only rarely does a newspaper refuse an ad on grounds of disagreement. If newspapers were as opinionated as they used to be in the days when they were competitive, public opinion would have long since acted against their unregulated monopoly.

Furthermore, newspapers are far from having a complete monopoly. Newspaper publishers, like cablecasters, argue that they are not a monopoly in an appropriately defined market. Even if there is only one newspaper in a town, there are many ways in which opinions get distributed in print. A handbill or periodical of opinion competes with a newspaper in the marketplace of ideas. News magazines and suburban papers also compete.

The Tornillo decision is not likely to be reconsidered. Newspapers are facing growing competition. Electronic information services and specialized national newspapers will erode still further their local monopoly. If such monopolies have not constrained open discussion in print up to now, they are even less likely to do so in the future.

Cablecasters claim that their situation is no different from that of the press, so they deserve exactly the same treatment. They argue that they too will maintain an open forum. Perhaps fifteen years hence one will be able to say that the cable industry saw the writing on the wall and behaved in a statesmanlike manner. Maybe it too will have voluntarily made channels available to all, even leasing channels to competitors. Maybe by then the newly emerging technology of a broadband ISDN on the telephone network will also have made it no longer sensible to talk of cable monopoly. But there is reason to doubt such expectations. The technical solution may come slowly, and the forecast of statesmanship is hardly supported by present behavior. Newspapermen come from a tradition of political combativeness and First Amendment principle; cablecasters come from the tradition of show business. Newspapers are an unregulated industry proud of their independence from the state; cable is a regulated franchised business. To look to the cable industry for such sensitivity to First Amendment considerations as to prevent the access issue from becoming intense is probably unrealistic.

There are economic reasons why radical surgery to separate carrier from content in cablecasting would not work in America. There is not now, nor will there be in the near future, the volume of carrier business needed for private cable businesses to expand in this way.[7] But given the temptation for a cable monopoly to stifle uses that do not interest it, and given the self-serving positions against requirements for channel leasing that the industry has taken, there is good reason for city governments, when franchising a cable monopoly, to require that at least leased access on a nondiscriminatory basis be provided. There are ways in which this can be done without destroying the economics of the systems. Nor does a leasing requirement in a franchise deny cablecasters their First Amendment rights.

Cable monopolies, owing to the physical problems of traversing the city's terrain, exist by grace of government franchise. Local newspapers are natural economic monopolies. This is a difference of a kind the courts have recognized.[8] The distinction can be stated in terms of resource constraint. Local newspaper monopolies arise from choices that consumers and suppliers make in the market, not from the existence of constraints that are so severe as to prevent the effective making of such market choices. Nonetheless, until the electronic media shake the present newspaper structure and bring readers into easy contact with competing news sources, local press monopolies will remain common. The paradox will continue of a monopolized print medium enjoying the freedom of the print tradition, while common carrier and regulated practices continue for electronic media, some of which operate under severe resource constraints and should therefore be obliged to provide access, and others of which do not.

The precise structure of common carrier regulation as embodied in the FCC's common carrier rules and the 1934 Communications Act is quite properly being questioned as burdensome. But the core of the common carrier concept, namely that a vendor with monopoly advantage in the market must provide access to customers without discrimination, remains often applicable to basic electronic carriers, as it was in the past to the mails.

The Policy Debate about Monopoly

Fear of monopoly has been at the core of most current communications policy debates and of most proposals to depart from the First Amendment tradition. "Monopoly" was the word used in 1927 by those attacking AT&T's attempt to set up a broadcasting common carrier. It is the word now being used by postal and telecommunications administrations in defense of their exclusive right to carry messages among the public. It is the word used to justify special restraints on AT&T.

Monopoly implies a single entity, but what is generally discussed is rather a matter of degree. A company of sufficient size to affect the market in which it operates is said in popular discourse to have some monopoly power. The television networks are frequently called monopolies, though there are three of them. The word "oligopoly" exists, but not in lay discussions. Furthermore, it describes only one of the ways in which partial monopoly power may exist. The very small publisher of a neighborhood shopping throwaway is most often a monopolist in the neighborhood but is in fierce competition for advertising with the city daily newspaper and thus has very little market power.

Market power is not identical with social or political power over communications, though they are closely related. The monopoly situations that are of concern for liberty are those where some resource needed to communicate is scarce enough that whoever owns it has considerable power over others who seek to use it. The economist's analysis focuses rather on power over other suppliers who compete in the market. A political analysis focuses instead on who gets to use the airwaves of a station that is licensed or who can send messages when a carrier controls the practical means for delivering them.

In rhetoric, the United States government favors diversity of voices and seeks to break up communications monopolies. The reality, however, is more ambiguous. Few monopolies exist from economic factors alone, and fewer still survive by private coercion alone. Mafiosos are not that strong. The force that preserves most monopoly privilege is the law. Some monopolies rest on patents, others on copyright, still others on franchises or licenses, some on property rights in unique locations, and many on regulatory policies that protect vested interests against assault. Most monopolies exist by grace of the police and the courts. From a social point of view some are desirable, others undesirable; but most would vanish in the absence of enforcement.

Antitrust policy, and thus most current debate over communications policy, has focused on the market-produced monopolies, for the monopolies that the government establishes by patents, copyrights, franchises, and laws are exceptions to the antitrust laws and so are perfectly legal. The government does not challenge them. When American government does grant a monopoly, its attitude is sometimes ambivalent. Monopoly grants are often designed to give a privilege and at the same time to limit it. Both copyrights and patents, for example, are for finite terms, require disclosure, and may not be used to keep a product off the market. They are monopolies intended in the end to promote rather than restrict access.

While the intent of regulation is often to provide some modest protection for the weak, the ultimate outcome is often more protection for the strong. American broadcasting regulation follows a policy of localism, that is, protecting local stations so a few superstations do not dominate the national air. This policy protects an oligopoly of broadcasters in every city. It gives them advantages not only in their own community but also against still bigger would-be national monopolists. Often regulation is thus used to give smaller companies some monopoly protection against larger ones. For decades neither AT&T nor Western Union were allowed to go into international telegraphy; it was reserved for four international record carriers which, it was believed, would be crushed if the domestic communication giants were allowed into the intercontinental business. For satellites too the policy of "open skies," by excluding AT&T, assured business in the formative years to a group of oligopolists.

The legal crutch that preserves weak companies is exculpated in the name of competition. If the crutch were removed, it is said, one more company would disappear, leaving fewer and larger contenders in control of the field. Thus in a normal communications environment there are little monopolies and big ones; each argues for the essentiality of its privilege, and each enjoys at most only a bit of monopoly.

Regulation, whatever its motives, tends to create these islands of segregated activity. Some firms are protected from others. Also, it is easier to control an activity when it is not mixed in with ones that are unregulated. A mix of competition and monopoly creates the possibility of cross-subsidization. Profits from a sheltered activity may be used to cut prices in competitive fields. The primary goal of antitrust policy in telecommunications has been to ensure that no one entity is simultaneously in both the naturally monopolistic portions of the phone business and in competitive markets.

At the same time, the goal of deregulation has been to free companies from the bonds of regulatory convenience and allow them to experiment in the market with the efficiencies of new technologies and joint products. In the United States this unleashing has been enjoyed by AT&T, but only AT&T without local operating companies. It has not been done for the postal service, nor is it likely to be done for such a tax-supported enterprise, though the same result may be achieved through private express carriers.

The postal system has an office in every neighborhood and delivery to every door. Historically, this made it seem a natural organization for also handling small parcels, which it now handles everywhere. It also appeared a natural organization for delivering telegrams. In countries where they are still delivered, this is done through the post office or else at enormous loss. Post offices also serve as convenient government field offices. In many countries a poor person's bank, plus the sale of money orders and sometimes insurance, is handled through the post office. The advantage of sharing joint costs among many functions of a distribution plant was perceived even at the birth of postal service, when monarchs got cheap mail service for themselves by allowing the recipients of their postal patent to carry the public's letters for a fee. Daily, to-the-door delivery could conceivably be made less of a fiscal burden if milk, eggs, newspapers, and mail were all handled together. It would be good economic policy for a postal service to get into other businesses in competition with haulers, telecommunications companies, banks, and dairies. Similarly telephone companies, which have a virtually universal billing system and a network for moving funds, may become billing services and, following that, credit organizations and financial intermediaries, or what are ordinarily called banks. By the same token banks may become communications carriers. Certainly computer and aerospace companies may find that they have the skills and facilities to offer transmission services. IBM, Comsat, and Aetna Life Insurance formed Satellite Business Systems to link computers and other business facilities.

Present American deregulatory policy encourages such competition. Any company can get into the game, except ones with a substantial monopoly position that could be used for anticompetitive practices. The popular cry now is to let the market determine which alternative vendors with their different joint costs, organization, and skills can efficiently provide each service. Increasingly, the government is allowing AT&T, Western Union, the international record carriers, the specialized common carriers, and the satellite companies onto each other's turf, and also for that matter banks, computer companies, railroads, and any company at all.

At least as important as ideology in causing communications deregulation has been technological change. The use of coaxial cable and of ever higher frequencies has eroded spectrum shortage. The introduction of microwave transmission in the 1930s eliminated the problem of right of way. Microwave frequencies, though not unlimited, were abundant enough to allow a substantial number of competing carriers. Satellite communication has reinforced this trend, for nothing prevents there being several competing satellite transmission organizations.

Deregulation, however, is a pragmatic policy. The argument made against regulations has been that they are inefficient and unnecessary, not improper. It holds that with converging technologies, the removal of controls will produce competition. Where this does not turn out to be the case, the deregulators are ready to step back in and regulate. But in the arena of speech and press we need also to consider other guidelines that have been left in the outfield in recent policy controversies—ones that recognize the preferred position of freedom of discourse.

Guidelines for Freedom

Difficult problems of press freedom, as well as of economics, arise at the intersections of regulatory models. When resource constraints are small and circumstances neatly fit the historical pattern of publishing, or when resource constraints are severe and circumstances fit the historical situation of a common carrier, then norms exist. The difficulties arise in situations that have elements of each. This was the problem in deciding about the broadcasting system in the 1920s; it is also a problem in the regulation of electronic networks today.

Regulators find it convenient to segregate activities and to keep each organization on its own turf. Much of regulatory law consists of specifications as to who may engage in what activities. Frequency allocations are made for particular uses; CBers or amateurs may not broadcast entertainment; public broadcast stations may not carry ads.[9] In the United States, AT&T and Western Union have been largely partitioned, with AT&T kept out of telex and telegraph traffic and Western Union kept out of voice. Deregulation loosens such restrictions and allows companies to move onto each other's turf. But some segmentation persists.

A price is paid for this rigid delimitation of turf, not only in efficiency and innovation but also in freedom of speech. The notion that government may specify which communications entity is allowed to participate in particular parts of the information industry's vertical flow is hard to reconcile with the First Amendment. To research and write, to print or orate, to publish and distribute, is everyone's right.

If government licensing of reporters, publishers, or printing presses is anathema, then so also should be the licensing of broadcasters and telecommunications carriers.

Yet the repeated argument has been made, which may be right or wrong in particular cases, that some degree of natural monopoly prevails in particular parts of the communications field. Whether because there were thought to be only 89 broadcasting frequencies, or because having more than one company digging up the streets was intolerable, or because the carrier that reached most persons was the one most worth joining, it seemed likely that a dominant organization would gain control of a communications resource that other citizens also needed. Under these circumstances the best solution seemed to be to define a monopoly's turf narrowly and to require those who had the monopoly to serve all comers without discrimination.

Since the institutions in such strategic positions are usually basic carriers of physical signals, one way to narrow their domain is to separate the carrier from content-related activities. But there are problems in doing this, in terms both of undercutting the economics of the business and, in America, of bending the Constitution. The unfortunate compromise that has often followed is to license and regulate the monopoly.

Such limited franchises have a way of being extended beyond their original rationale. Enfranchised monopolies that at one time are thought simply to reflect in an orderly fashion the natural realities of the market, and are indeed intended to restrict monopoly, get converted into matters of right. Stations and carriers that are licensed simply to ensure good service by carefully selected organizations when monopoly seems inevitable come to see themselves, and to be seen, as having a vested right in their franchise. Regulatory powers assumed by the government to cope with monopoly also acquire a life of their own.

This faces the communications field with a dilemma. Not all parts of the communications system fit well under the preferred print model. Bottlenecks do exist where there are severe resource constraints. And the regulations that in those situations seem to be required have an insidious bent. They acquire legitimacy; they outlive their need; they tend to spread. The camel's nose is under the tent.

Yet when there is severe scarcity, there is an unavoidable need to regulate access. Caught in the tension between the tradition of freedom and the need for some controls, the communications system then tends to become a mix of uncontrolled and common carrier elements—of anarchy, of property, and of enfranchised services. A set of principles must be understood if communications in the electronic era are to hold as fully as possible to the terms of the First Amendment. The technology does not make this hard. Confusion about principles may.

The first principle is that the First Amendment applies fully to all media. It applies to the function of communication, not just to the media that existed in the eighteenth century. It applies to the electronic media as much as to the print ones.

Second, anyone may publish at will. The core of the First Amendment is that government may not prohibit anyone from publishing. There may be no licensing, no scrutiny of who may produce or sell publications or information in any form.

Third, enforcement of the law must be after the fact, not by prior restraint. In the history of communications law this principle has been fundamental. Libel, obscenity, and eavesdropping are punishable, but prior review is anathema. In the electronic media this has not been so, but it should be. Traffic controls may be needed in cases where only one communicator can function at a particular place at a particular time, such as street meetings or use of radio frequencies, but this limited authority over time and place is not the same as power to choose or refuse to issue a license.

Fourth, regulation is a last recourse. In a free society, the burden of proof is for the least possible regulation of communication. If possible, treat a communications situation as free for all rather than as subject to property claims and a market. If resource constraints make this impossible, treat the situation as a free market rather than as a common carrier. But if resources for communication are truly monopolistic, use common carrier regulation rather than direct regulation or public ownership. Common carriage is a default solution when all must share a resource in order to speak or publish.

Under common law in the nineteenth century, vendors could not be made common carriers against their will.[10] If they offered a service to the general public, it had to be without discrimination, but if they chose to serve a limited clientele, that was their right. This philosophy applies well to publishing. One would not require the Roman Catholic *Pilot* to carry ads for birth control or a trade union magazine to carry ads against the closed shop. But these cases assume that diverse magazines exist. A dilemma arises when there is a monopoly medium, as when a monopoly newspaper in a town refuses ads to one party and carries them for another.

In the world of electronic communications some but not all of the basic physical carriers, and only those, seem likely to continue to have significant monopoly power. It is hard to imagine a value-added network having the dominance in a community that a local newspaper has today. Even now the communications monopolies that exist without privileged enforcement by the state are rare. Even basic physical conduits become monopolies precisely because they cannot exist without public favors. They need permissions that only the state can grant. These favors, be they franchises to dig up the city streets or spectrum to transmit through the air, may properly be given to those who choose to serve as common carriers. This is not a new idea. In 1866 telegraph companies were given the right to string wires at will along post roads and across public lands, but only if they became common carriers. Where monopoly exists by public favor, public access is a reasonable condition.

Fifth, interconnection among common carriers may be required. The basic principle of common carriage, namely that all must be served without discrimination, implies that carriers accept interconnection from each other. This principle, established in the days of the telegraph, is incorporated in the 1982 AT&T consent decree. All long-distance carriers have a right to connect to all local phone companies. That is the 1980s outgrowth of the 1968 Carterphone decision which required AT&T to interconnect with an independent radio-telephone service.[11] Universal interconnection implies both adherence to technical standards, without which interconnection can be difficult, and a firm recognition of the right to interconnect.

Carriers may sometimes raise valid objections to interconnection. Some will wish to use novel technologies that are incompatible with generally accepted standards, claiming that they are thereby advancing the state of the art. Also, when they handle highly sensitive traffic, such as funds transfers or intracompany data, they may not wish to be common carriers and bear the risks of having outsiders on their system. Such arguments are often valid, though they may also be used to lock a group of customers out of using the carrier.

An argument in favor of general interconnectivity is that it facilitates market entry by new or small carriers. It also makes universal service easier. It may even be useful for national security, since a highly redundant system is less likely to be brought down. In short, there are conflicting considerations that must be balanced. As a policy, the requirement of interconnection is a reasonable part of a common carriage system.

Sixth, recipients of privilege may be subject to disclosure. The enforcement of nondiscrimination depends critically on information. Without control of accounting methods, regulatory commissions are lost in a swamp. I once asked the head of the Common Carrier Bureau of the FCC what he would ask for if he could rub Aladdin's lamp. "Revelatory books" was his reply.

Yet American lawmakers, who have imposed far more oppressive and dubious kinds of regulation, such as exit, entry, and tariff controls, have never pushed the mild requirement for visibility. Apart from requiring accounts, legislators have been highly considerate of proprietary information. A firm that enjoys the monopoly privileges which lead to being a common carrier should perhaps forgo, like government, some privileges of privacy. Unbundled rates for cable leasing, for example, help reveal who is being charged for what. Disclosure is not a new idea. Patents and copyrights are privileges won only by making their object public. The same principle might well apply to action under franchises too.

Seventh, privileges may have time limits. Patents and copyrights are for finite periods, and then the right expires. Radio and television licenses and cable franchises, though also for fixed periods, are typically renewable. Some monopoly privileges that broadcasters and cablecasters have in their licenses could expire after a fixed period. This is a way to favor infant industries but limit their privileges when they become giants.

Eighth, the government and common carriers should be blind to circuit use. What the facility is used for is not their concern. There may be some broad categories of use. Emergency communications often have priority. Special press rates for telegraph have been permitted, though their legality in the United States has been questioned.[12] But in general, control of the conduit may not become a means for controlling content. What customers transmit on the carrier is no affair of the carrier.

Ninth, bottlenecks should not be used to extend control. Rules on undeliverable mail have been used to control obscene content. Cablecasting, in which there is no spectrum shortage, has been regulated by the FCC as ancillary to broadcasting. Telegraph companies have sought to control news services, and cable franchisees have sought to control the programs on the cable. Under the First Amendment, no government imposition on a carrier should pass muster if it is motivated by concerns beyond common carriage, any more than the carrier should be allowed to use its service to control its customers.

Tenth, and finally, for electronic publishing, copyright enforcement must be adapted to the technology. This exceptional control on communication is specifically allowed by the Constitution as a means of aiding dissemination, not restricting it. Copyright is temporary and requires publication. It was designed for the specific technology of the printing press. It is in its present form ill adapted to the new technologies. The objective of copyright is beyond dispute. Intellectual effort needs compensation. Without it, effort will wither. But to apply a print scheme of compensation to the fluid dialogue of interactive electronic publishing will not succeed. Given modern technologies, there is no conceivable way that individual copies can be effectively protected from reproduction when they are already either on a sheet of paper or in a computer's memory. The task is to design new forms of market organization that will provide compensation and at the same time reflect the character of the new technology.

The question boils down to what users at a computer terminal will pay for. For one thing, they will pay for a continuing relationship, as they will continue to need maintenance. It may be easy to pirate a single program or some facts from a data base by copying from a friend of a friend of a friend who once bought it. But to get help in adapting it or to get add-on versions or current data, one might pay a fee as a tender for future relations. The magazine subscription model is closer to the kind of charging system that will work for electronic publishing than is the one-time book purchase with a royalty included.

A workable copyright system is never enacted by law alone. Rather it evolves as a social system, which may be bolstered by law. The book and music royalty systems that now exist are very different from each other, reflecting the different structures of the industries. What the law does is to put sanctions behind what the parties already consider right. So too with electronic publishing on computer networks, a normative system must grow out of actual patterns of work. The law may then lend support to those norms.

If language were as fluid as the facts it represents, one would talk in the electronic era of serviceright, not copyright. But as language is used, old words are kept regardless of their derivation, and their meanings are changed. In the seventeenth century reproducing a text by printing was a complex operation that could be monitored. Once the text was printed on paper, however, it required no further servicing, and no one could keep track of it as it passed from reader to reader. In the electronic era copying may become trivially easy at the work stations people use. But both the hardware and the software in which the text is embodied require updating and maintenance. In ways that cannot yet be precisely identified, the bottleneck for effective monitoring and charging is migrating from reproduction to the continuing service function.

Not only in copyright but in all other issues of communications policy, the courts and legislatures will have to respond to a new and puzzling technology. The experience of how the American courts have dealt with new nonprint media over the past hundred years is cause for alarm. Forty years ago Zechariah Chafee noted how differently the courts treated the print media from newer ones: "Newspapers, books, pamphlets, and large meetings were for many centuries the only means of public discussion, so that the need for their protection has long been generally realized. On the other hand, when additional methods for spreading facts and ideas were introduced or greatly improved by modern inventions, writers and judges had not got into the habit of being solicitous about guarding their freedom. And so we have tolerated censorship of the mails, the importation of foreign books, the stage, the motion picture, and the radio."[13] With the still newer electronic media the problem is compounded. A long series of precedents, each based on the last and treating clumsy new technologies in their early forms as specialized business machines, has led to a scholastic set of distinctions that no longer correspond to reality. As new technologies have acquired the functions of the press, they have not acquired the rights of the press. On print, no special excise taxes may be applied; yet every month people pay a special tax on their telephone bill, which would seem hardly different in principle from the old English taxes on newspapers. On print, the court continues to exercise special vigilance for the preferred position of the First Amendment; but other considerations of regulatory convenience and policy are given a preferred position in the common carrier and electronic domains.

Since the lines between publishing, broadcasting, and the telephone network are now being broken, the question arises as to which of these three models will dominate public policy regarding the new media. There is bound to be debate, with sharp divisions between conflicting interests. Will public interest regulation, such as the FCC applies, begin to extend over the conduct of the print media as they increasingly use regulated electronic means of dissemination? Or will concern for the traditional notion of a free press lead to finding ways to free the broadcast media and carriers from the regulation and content-related requirements under which they now operate?

Electronic media, as they are coming to be, are dispersed in use and abundant in supply. They allow for more knowledge, easier access, and freer speech than were ever enjoyed before. They fit the free practices of print. The characteristics of media shape what is done with them, so one might anticipate that these technologies of freedom will overwhelm all attempts to control them. Technology, however, shapes the structure of the battle, but not every outcome. While the printing press was without doubt the foundation of modern democracy, the response to the flood of publishing that it brought forth has been censorship as often as press freedom. In some times and places the even more capacious new media will open wider the floodgates for discourse, but in other times and places, in fear of that flood, attempts will be made to shut the gates.

The easy access, low cost, and distributed intelligence of modern means of communication are a prime reason for hope. The democratic impulse to regulate evils, as Tocqueville warned, is ironically a reason for worry. Lack of technical grasp by policy makers and their propensity to solve problems of conflict, privacy, intellectual property, and monopoly by accustomed bureaucratic routines are the main reasons for concern. But as long as the First Amendment stands, backed by courts which take it seriously, the loss of liberty is not foreordained. The commitment of American culture to pluralism and individual rights is reason for optimism, as is the pliancy and profusion of electronic technology.

Notes

1. A Shadow Darkens

1. Charles D. Ferris, quoted in *The Report*, Nov. 14, 1980, p. 11.

2. Nina McCain, *Boston Globe*, Field News Service, Aug. 31, 1980.

3. "Press Freedom—A Continuing Struggle," speech to Associated Press Broadcasters Convention, June 6, 1980; *New York Times*, July 7, 1980, sec. B, p. 3. See also Judge David Bazelon, "The First Amendment and the New Media," *Federal Communications Law Journal* 31.2 (Spring 1979); Bazelon, "The First Amendment's Second Chance," *Channels*, Feb.-Mar. 1982, pp. 16–17; Charles Jackson, Harry M. Shooshan, and Jane L. Wilson, *Newspapers and Videotex: How Free a Press?* (St. Petersburg, Fla.: Modern Media Institute, 1982); John Wicklein, *Electronic Nightmare—The New Communications and Freedom* (New York: Viking, 1981); and an early statement by this author, Ithiel de Sola Pool, "From Gutenberg to Electronics: Implications for the First Amendment," *The Key Reporter* 43.3 (Spring 1978).

4. Alexis de Tocqueville, *Democracy in America* (1840, reprint New York: Knopf, 1945), II, 316–318.

5. Daniel Bell, *The Cultural Contradictions of Capitalism* (New York: Basic Books, 1975), p. 10.

6. Mark U. Porat and Michael R. Rubin, *The Information Economy*, 9 vols. (Washington, D.C.: Government Printing Office, 1977); Organization for Economic Cooperation and Development, *Information Activities, Electronics and Telecommunications Technologies: Impact on Employment, Growth and Trade* (Paris: 1981).

2. Printing and the Evolution of a Free Press

1. Harold A. Innis, *The Bias of Communication* (Toronto: University of Toronto Press, 1951), pp. 18–19; Lynn T. White, Jr., "Technology Assessment from the Stance of a Medieval Historian," *Technological Forecasting and Social Change* 6 (1974): 366.

2. Label in Gutenberg Museum, Mainz.

3. Felix and Marie Keesing, *Elite Communications in Samoa: A Study of Leadership* (Stanford: Stanford University Press, 1956).

4. See Samuel Popkin, *The Rational Peasant* (Berkeley: University of California Press, 1979), pp. 106–107, for the interplay of consensus and conflict in the village council system of Viet Nam. The effective right of influential citizens to veto decisions inhibited discussion. The possibility of veto by single objectors is likely to discourage provocative statements.

5. For anthropological discussions of the concept of freedom, see Bronislaw Malinowski, *Freedom and Civilization* (New York: Ray Publishers, 1944); Vilhjalmur Stefansson in Ruth Nanda Anshen, *Freedom, Its Meaning* (New York: Harcourt, Brace, 1940); Margaret Mead, *Cooperation and Competition among Primitive Peoples* (New York: McGraw-Hill, 1937). Malinowski uses freedom as a holistic concept of human welfare and so is hard to interpret regarding civil liberties. Stefansson argues that Eskimo society was freer than modern ones, as evidenced by the absence of chiefs, prisons, and floggings. He perceives the elaborate Eskimo system of taboos as like advice we accept from physicians and pay for. Of the thirteen cultures Mead studied, the three societies that seemed freest—the Arapesh, Eskimo, and Onbwan—were individualistic in the sense that individuals strove toward their goals without reference to others. However, the four freest societies were at the lowest level of development, and the six least free societies included the four richest.

6. Denys Hay, Introduction to John Carter and Percy Muir, *Printing and the Mind of Man* (New York: Holt, Rinehart and Winston, 1967), p. 742; Elizabeth Eisenstein, "Some Conjectures about the Impact of Printing on Western Society and Thought: A Preliminary Report," *Journal of Modern History* 40.1 (1968): 3, which is a trenchant summary of her classic, *The Printing Press as an Agent of Change*, 2 vols. (Cambridge, Eng.: Cambridge University Press, 1979).

7. "An Episode in the History of Social Research: A Memoir," in Donald Fleming and Bernard Bailyn, eds., *The Intellectual Migration: Europe and America, 1830–1960* (Cambridge: Harvard University Press, 1969), p. 320.

8. Eisenstein, "Some Conjectures," p. 22.

9. This figure for printers is estimated from the Venice statistic above. It is not clear whether helpers were included in that figure. If not, the figure might be as few as two volumes a week instead of one a day, but the conclusion is unaffected.

10. Eisenstein, "Some Conjectures," p. 4.

11. Eisenstein, "Some Conjectures," p. 20.

12. Eisenstein, "Some Conjectures," p. 36.

13. Edward C. Caldwell, "Censorship of Radio Programs," *Journal of Radio Law* 1.3 (Oct. 1931): 444.

14. Decree of 1586, quoted in John Shelton Lawrence and Bernard Timberg, *Fair Use and Free Inquiry* (Norwood, N.J.: Ablex, 1980), pp. 4–5; *Encyclopaedia Britannica* (1958), "Printing."

15. Eisenstein, "Some Conjectures," p. 52.

16. *Britannica*, "Periodicals."

17. Commonwealth v. Blanding, 3 Pick 304, 15 Am. Dec. 214.

18. Grossjean v. American Press Co., 297 US 233, 80 L.ed. 660, 56 S.Ct. 444.

19. New York Times v. Sullivan, 376 US 254, 11 L.ed. 2nd 686, 84 S.Ct. 710.

20. The full text reads: "Congress shall make no law respecting an establishment of religion, or prohibiting the free exercise thereof; or abridging the freedom of speech or of the press; or the right of the people peaceably to assemble, and to petition the government for a redress of grievances."

21. The word "copyright" first appears in 1697 in Blackstone's *Commentaries*. "However, the concept of copyright goes back much further than Blackstone. The right only began to assume importance when the invention of printing made the multiplication of 'copies' of a work infinitely quicker and cheaper than the painstaking products of monkish scribes, as well as appreciably more accurate than the compositions of most professional scriveners." Ian Parsons, "Copyright and Society," in Asa Briggs, ed., *Essays in the History of Publishing* (London: Longman, 1974), pp. 31, 33–34.

22. White Smith v. Apollo 209 US 1, 52 L.ed. 655, 28 S.Ct. 319 (1908). See also Goldstein v. California, 412 US 546, 37 L.ed. 163, 93 S.Ct. 2303 (1973), on sound recordings.

23. Art. 1, sec. 8, para. 7.

24. See Wesley E. Rich, *The History of the United States Post Office to the Year 1829* (Cambridge: Harvard University Press, 1924).

25. Today concessionary mail rates exist in twelve of thirteen European countries studied (not Ireland), and concessionary telephone and telegraph rates exist in eleven (not Britain or Ireland). Anthony Smith, *Subsidies and the Press in Europe* (London: PEP, 1977).

26. See Max Horkheimer and Theodor W. Adorno, *Dialectic of Enlightenment* (1947, reprint New York: Seabury Press, 1972).

27. After the tax was reduced to one penny in 1836, the number of copies on which the tax was paid grew from 39 million to 122 million by 1854, and with the abolition of the tax in 1855 growth surged further. *Britannica*, "Newspapers."

28. Pauline Wingate, "Newsprint: From Rags to Riches—and Back Again," in Anthony Smith, ed., *Newspapers and Democracy* (Cambridge: MIT Press, 1980), p. 67.

29. Wingate, "Newsprint," p. 65.

30. David C. Smith, "Wood Pulp and Newspapers, 1867–1900," *Business History Review* 3 (Autumn 1964): 328–345.

31. Daniel J. Boorstin, *The Americans: The Democratic Experience* (New York: Random House, 1958).

32. Ithiel de Sola Pool *et al.*, *Communications Flows in the U.S.: A Census with Comparisons to Japan*, forthcoming.

33. Flow census data comparable to those for the United States exist

only for Japan. For other countries only circulation and publication figures are available. Newspaper circulation per thousand of population is higher in many advanced countries than in the United States. But, American newspapers are far thicker, so American figures tend to lead in number of words delivered. In 1975 in Japan, for example, newspaper circulation per thousand was 229 and in the United States only 134, but owing to the size of American papers, US newsprint consumption in 1977 was 10.2 million tons and in Japan only 2.29 million tons. Wingate, "Newsprint," p. 68. As incomes elsewhere rise to American levels, the print volume may be expected to show some of the same saturation features as in the United States.

3. Electronics Takes Command

1. Innis, *The Bias of Communication.*
2. Alvin F. Harlow, *Old Wires and New Waves* (New York: D. Appleton-Century, 1936), pp. 40–43.
3. Punch, XI, 253. See also XXXV (1858), 254. "House" refers to a home, not the inventor Royal E. House, but perhaps the pun was intended.
4. To protect its Gold and Stock business, Western Union in 1876–1879 went into competition with the Bell system in telephone development, basing its try in part on the Edison and Elisha Grey telephone patents, but the Bell patents won in court.
5. Sidney H. Aronson, "Bell's Electrical Toy," in Ithiel de Sola Pool, *The Social Impact of the Telephone* (Cambridge: MIT Press, 1977), p. 15. The widely believed story is apocryphal that Orton, the president of Western Union, dismissed the invention with the statement, "What use could this company make of an electrical toy?" He fully realized the importance of the invention but misjudged his ability to proceed in telephony without the Bell patent. Failing in the effort to compete using the Grey and Edison patents, Western Union reached an agreement with the Bell people in 1879 providing for patent exchanges and royalty payments and promising that Western Union would stay out of telephony and the phone company out of telegraphy.
6. The phrase "grand system" was among President Vail's key slogans for the phone company, as was the phrase "universal service."
7. This was far from a trivial matter, for next to copper, trees for poles were the most critically scarce resource for the phone industry. As time passed and the phone system grew, it became apparent that the demand for such trees would exhaust the supply, which was an incentive for seeking alternatives, such as underground cables or cement poles. *Telephony* 10.1 (July 1905): 51; Burton J. Hendrick, "Telephones for the Millions," *McClure's Magazine* 44 (Nov. 1914): 55. Joint use of poles for telephony and telegraphy was obviously economic, whether or not the wires were used in common.
8. Many of the circuits used by Western Union have been leased from AT&T. It is a cost judgment by Western Union whether to pay to install its

own physical cable or microwave or to lease a circuit. This is one more instance of the convergence of the two systems.

9. The early sound movie equipment, called Vitaphone, was produced by a subsidiary of Bell's, and when in the depression of the 1930s motion picture companies were going bankrupt, the Bell System sought to protect its investment in sound cinematography by becoming a main financial backer of movies. So aggressive research led to a convergence of the telephone and movie industries. Under the pressure of antitrust policy, however, this relation did not last.

10. Pool, *The Social Impact of the Telephone*, pp. 40–65. See also Aronson, "Bell's Electrical Toy."

11. Security considerations also undoubtedly entered in, though there was no Radio Liberty broadcasting by shortwave at the time.

12. *The Independent* 73 (Oct. 17, 1912): 886–891.

13. See John Ward, "Present and Probable CATV/Broadband Communication Technology," in Ithiel de Sola Pool, ed., *Talking Back* (Cambridge: MIT Press, 1973), pp. 139–186.

14. See Otto Riegle, *Mobilizing for Chaos* (New Haven: Yale University Press, 1934).

15. W. P. Banning, *Commercial Broadcast Pioneer: The WEAF Experiment, 1922–1926* (Cambridge: Harvard University Press, 1946); Erik Barnouw, *A History of Broadcasting in the United States* (New York: Oxford University Press, 1966), Vol. I, *A Tower in Babel*, pp. 105–114.

16. High-frequency (HF) or short waves, which are actually intermediate between medium waves and VHF, also go in fairly straight lines but are reflected back to earth by the ionosphere, so they too can be received at much greater distances than medium waves. They are, therefore, the ones used for international broadcasting.

17. In the future even shorter waves will be put to similar uses. One problem with still shorter waves, like light, is that they are blocked by rain and fog. One advantage is that there is a very large number of such frequencies. Technological ways of bypassing some of the difficulties can be found.

18. The 1982 AT&T consent decree makes this a more economically attractive option than in the past because calls that do not use the local operating company evade the access charges that such a company will impose.

19. A common example of multiplexing of broadcast signals is sending Muzak-type service to subscribers on one side lobe of an ordinary FM radio station's transmissions.

20. Paul F. Lazarsfeld, *Radio and the Printed Page* (New York: Duell, Sloan and Pearce, 1940).

21. Hilda Himmelweit, *TV and the Child* (London: Oxford University Press, 1958).

22. The growth of radio news and particularly of shortwave reception compelled the Communist Party of the Soviet Union in 1963 to redefine the respective role of Pravda and radio broadcasters. Until then, all media, including radio newscasters, had had to wait for the Pravda story to learn how to treat a news item. This delay gave foreign broadcasters an edge on up-to-the-minute news. So the instructions were changed, giving radio newscasters the mission of first announcement of news, while Pravda was given the mission of its fuller interpretation. Ithiel de Sola Pool and Wilbur Schramm, eds., *Handbook of Communication* (Chicago: Rand McNally, 1974), p. 482.

23. Benjamin M. Compaine, *Who Owns the Media*, 2nd ed. (White Plains, N.Y.: Knowledge Industries Publications, 1982), pp. 30,38,39. See also Benjamin M. Compaine, *The Newspaper Industry in the 1980s* (White Plains, N.Y.: Knowledge Industry Publications, 1980), pp. 86, 88; James Rosse, Bruce M. Owen, and James Dertouzos, "Trends in the Daily Newspaper Industry, 1923–1973, Studies in Industry Economics no. 57 (Stanford University: Department of Economics, 1975), p. 30; Bruce M. Owen, *Economics and Freedom of Expression* (Cambridge: Ballinger, 1975), p. 81.

24. Compaine, *The Newspaper Industry*, p. 39. Magazine and book publishing also remains highly competitive. The number of magazines rose by 56% during 1950–1981, and during 1947–1977 the percentage of the business accounted for by the top four companies declined from 34% to 22% and by the top eight companies declined from 43% to 35%. The total number of books published annually during 1946–1980 grew at a compound annual rate of 5.2%, and during 1925–1977 the percentage of shipments accounted for by the top four companies fell from 20% to 17% and by the top eight companies stayed at 30%. Compaine, *Who Owns the Media*, pp. 100, 102, 194, 158.

25. Compaine, *The Newspaper Industry*, p. 88. The top 10% had 61.3%.

26. Anthony Smith, *Goodbye Gutenberg* (New York: Oxford University Press, 1980); Smith, *Newspapers and Democracy*, ch. 1.

27. Compaine, *Who Owns the Media*, pp. 48, 318.

28. Compaine, *Who Owns the Media*, pp. 53, 189, 326, 328, 338, 389, 396, 421.

29. Chart by David Allen, revised from *Annual Report*, Research Program on Communications Policy (Cambridge: MIT, 1982), p. 12.

30. FCC, Second Report and Order on Docket 18110, Jan. 31, 1975, sustained by Supreme Court in FCC v. National Citizens' Committee for Broadcasting, 98 S. Ct. 2096, 56 L.Ed 2nd 697, 436 US 775 (1978).

31. Compaine, *The Newspaper Industry*, p. 104.

32. A similar policy is evolving in Canada, where in the last two or three years several newspapers have closed their doors, leaving local newspaper monopolies behind. A Commission on Newspapers, the Kent Commission, expressed concern at the prospect of local monopoly not only in the press but across all media. In 1982 the Cabinet instructed the Canadian Radio and Television Commission (the equivalent of the FCC) to limit cross-owner-

ship. The commission (CRTC) will be developing policy to do so as it reviews renewal applications from broadcasting stations that are publisher owned.

33. Compaine, *Who Owns the Media*, p. 386. By FCC regulation these cable systems have to be located elsewhere than where the broadcasting and phone companies have their ordinary operations.

4. The First Amendment and Print Media

1. Justice Harlan argued in 1906 in a dissent in Patterson v. Colorado, "If the rights of free speech and of a free press are in their essence, attributes of national citizenship, as I think they are, then neither Congress nor any state, since the adoption of the 14th Amendment, can by legislative enactments or by judicial action impair or abridge them." 205 US 454, 51 L.ed. 879, 27 S.Ct. 556. Justice Holmes in the majority decision recognized the issue Harlan had raised but asserted that court precedent had not yet extended the First Amendment to the states. The Supreme Court did so in 1924 in Gitlow v. New York 268 US 652, 69 L.ed. 1138. By 1930 Justice Hughes could declare, "It is no longer open to doubt that the liberty of the press and of speech is within the liberty safeguarded by the due process clause of the 14th Amendment from invasion by state action." Near v. Minnesota, 283 US 697, 75 L.ed. 1357, 51 S.Ct. 625 (1930). In 1952 Justice Jackson tried to reopen the issue. Beauharnais v. Illinois, 343 US 250, 96 L.ed. 919, 72 S.Ct. 752. But the Court continued regularly to assert that the Fourteenth Amendment incorporates the First. See Burstyn v. Wilson, 343 US 495, 96 L.ed. 1098, 72 S.Ct. 777.

2. New York Times Co. v. Sullivan, 376 US 254, 11 L.ed. 2d 686, 84 S.Ct. 710.

3. Smith v. California, 361 US 147, 4 L.ed. 2d 205, 80 S.Ct. 215 in concurrence.

4. Thomas Jefferson, *Works*, ed. Paul L. Ford (New York: G.P. Putnam, 1904–1905), pp. 464–465.

5. James Madison, *Writings*, ed. Gaillard Hunt (New York: G.P. Putnam, 1906), p. 391, quoted in Smith v. California.

6. Samuel Stouffer, *Communism, Conformity and Civil Liberties* (New York: Doubleday, 1955). See also Clyde Z. Nunn, Harry J. Crockett, Jr., and J. Allen Williams, Jr., *Tolerance of Nonconformity* (San Francisco: Jossey-Bass, 1978); James A. Davis, "Communism, Conformity, Cohorts, and Categories: American Tolerance," *American Journal of Sociology* 8 (1975): 491–513; Edward N. Muller, Pertti Personen, and Thomas Jukam, "Support for Freedom of Assembly in Western Democracies," *European Journal of Political Research* 8 (1980): 265–288; John L. Sullivan, James Pierson, and George E. Marcus, "A Reconceptualization of Political Tolerance: Illusory Increases, 1950s–1970s," *American Political Science Review* 73 (1979): 781–794; James L. Gibson and Richard D. Bingham, "On the Conceptualization and Measure-

ment of Political Tolerance," *American Political Science Review* 76 (1982): 603–620.

7. Joseph Story, *Commentaries on the Constitution*, quoted by Justice Butler in a dissent in Near v. Minnesota, 283 US 697, 75 L.ed. 1357, 51 S.Ct. 625 (1930).

8. Commonwealth v. Blanding, 3 Pick 304, 15 Am. Dec. 214.

9. Robertson v. Baldwin, 165 US 275, 281, 41 L.ed. 715, 717, 17 S.Ct. 326.

10. Dennis v. US, 341 US 494, 95 L.ed. 1137, 71 S.Ct. 857.

11. 283 US 697, 75 L.ed. 1357, 51 S.Ct. 625.

12. 4 Blackstone Commentaries, quoted in Near v. Minnesota.

13. For a historical treatment of the real-life story of the case, see Fred Friendly, *Minnesota Rag: The Dramatic Story of the Landmark Supreme Court Case That Gave New Meaning to Freedom of the Press* (New York: Random House, 1981).

14. Lovell v. Griffin, 303 US 444, 82 L.ed. 949, 58 S.Ct. 666.

15. Thomas v. Collins, 323 US 516, 89 L.ed. 430, 65 S.Ct. 315.

16. Staub v. Baxley, 355 US 313, 2 L.ed. 2d 302, 78 S.Ct. 277.

17. Bantam Books v. Sullivan, 372 US 58, 9 L.ed. 2d 584, 83 S.Ct. 631.

18. New York Times v. US, 403 US 713, 29 L.ed. 2d 822, 91 S.Ct. 2140.

19. Thus the Supreme Court's rejection of prior restraint, while historically strong, has not been absolute. In 1918 Justice Holmes, developing the "clear and present danger" test, held that anticonscription propaganda in wartime could be suppressed. Schenck v. US, 249 US 47, 63 L.ed. 470, 39 S.Ct. 247. Hughes in Near v. Minnesota noted draft-resistance in wartime as an example of extreme abuse that might be restrained. The furthest the Court has gone in allowing prior restraint is in the troubled area of obscenity, where the Court oscillates with changes in its composition and in the mores of the times. In 1952 the Court barred precensorship of movies, ruling that the constitutional protection of free speech applies to cinema, though not necessarily under the same rules as for other forms of expression. Burstyn v. Wilson, 343 US 495, 96 L.ed. 1098, 72 S.Ct 777. Somehow stretching that last qualification, the Court nine years later, in a five to four decision, sustained a Chicago ordinance requiring the submission of a print of a film to the censors before its showing. The film distributor had simply declined to submit the print for prior review. The Court excluded from its consideration how the Chicago authorities might proceed constitutionally to deal with a film, limiting the judgment to the question of whether they might insist on the advance submission of a print. The authorities, the Court held, might legitimately conclude that movies have so much greater power for evil by obscenity than the printed word that a prior review is needed, albeit conducted under standards consistent with the First Amendment. Times Film Corp. v. Chicago, 365 US 43, 5 L.ed. 2d 403, 81 S.Ct. 391. When in 1965 the Court finally dealt with the substance of a film censorship law, it laid down such stringent procedural requirements of prompt action

and judicial review as to make the normal bureaucratic operation of a censor virtually impossible. Freedman v. Maryland, 380 US 51, 13 L.ed. 2d 649, 85 S.Ct. 734.

20. US v. Progressive, Inc., 467 F. Supp. 990 (W.D. Wis. 1979).

21. 249 US 47, 63 L.ed. 470, 39 S.Ct. 247.

22. American Communications Association v. Douds, 339 US 382, 94 L.ed. 925. See also Dennis v. US, 341 US 494, 95 L.ed. 1137, 71 S.Ct. 857; Barenblatt v. US, 360 US 109, 3 L.ed. 2d 1115, 79 S.Ct. 1081; Uphaus v. Wyman, 360 US 72, 3 L.ed. 2d 1090, 79 S.Ct. 1040; Wilkinson v. US, 365 US 399, 5 L.ed. 2d 633, 81 S.Ct. 567; Braden v. US, 365 US 431, 5 L.ed. 2nd 653, 81 S.Ct. 584; Communist Party of the USA v. Subversive Activities Control Board, 367 US 1, 6 L.ed. 2nd 625, 81 S.Ct. 1357; Scales v. US, 367 US 203, 6 L.ed. 2nd 782, 81 S.Ct. 1469. The anti-Communist cases illustrate the element of truth in the old aphorism that the Court follows the election returns. With a declining public perception of a Communist menace, partly as a result of the decline of the American Communist movement, and partly with the slowing of the surge of world Communist expansion after the 1950s, the Court, starting in 1962, affirmed the First Amendment rights of Communists and of those investigated as possible Communists. See e.g. Russell v. US, 369 US 749, 8 L.ed. 2d 240, 82 S.Ct. 1038 (1962); Bagett v. Bullitt, 377 US 360, 12 L.ed. 2d 377, 84 S.Ct. 1316 (1964); Aptheker v. Secretary of State, 378 US 500, 12 L.ed. 992 84 S.Ct. 1659 (1964); Dombrowski v. Pfister, 380 US 479, 14 L.ed. 2d 22, 85 S.Ct. 1116 (1965); Lamont v. Postmaster General, 381 US 301, 14 L.ed. 2d 398, 85 S.Ct. 1493 (1965); Brandenburg v. Ohio, 395 US 444, 23 L.ed. 2d 430, 89 S.Ct. 1827 (1969); Communist Party of Indiana v. Whitcomb, 414 US 441, 38 L.ed. 2d 635, 94 S.Ct. 656 (1974).

23. Schenck v. US, 249 US 47, 63 L.ed. 470, 39 S.Ct. 247. See also Frohwerk v. US, 249 US 204, 63 L.ed. 561, 39 S.Ct. 249; Debs v. US, 249 US 211, 63 L.ed. 566, 39 S.Ct. 252; Abrams v. US, 250 US 616, 63 L.ed. 1173, 40 S.Ct. 17; Schaeffer v. US, 251 US 466, 64 L.ed. 360, 40 S.Ct. 259; Pierce v. US, 252 US 239, 64 L.ed. 542, 40 S.Ct. 205.

24. Feiner v. New York, 340 US 315, 95 L.ed. 295, 71 S.Ct. 303. See also Kunz v. New York, 340 US 290, 95 L.ed. 280, 71 S.Ct. 312; Chaplinsky v. New Hampshire, 315 US 568, 86 L.ed. 1031, 62 S.Ct. 766. But for cases where the Court found the police judgment unjustified, see Terminiello v. Chicago, 337 US 1, 93 L.ed. 1131, 69 S.Ct. 894; Edwards v. South Carolina, 372 US 229, 9 L.ed. 2d 697, 83 S.Ct. 680.

25. Bridges v. California, 314 US 252, 86 L.ed. 192, 62 S.Ct. 190. See also Thomas v. Collins, 323 US 516, 89 L.ed. 430, 65 S.Ct. 315; Abrams v. US, Holmes and Brandeis dissent, 250 US 616, 630, 631, 63 L.ed. 1173, 1180, 40 S.Ct. 17; Whitney v. California, Brandeis and Holmes concurring, 274 US 357, 376, 71 L.ed. 1095, 1106, 47 S.Ct. 641.

26. Concurrence in Pennekamp v. Florida, 328 US 331, 353, 90 L.ed. 1295, 1307, 66 S.Ct. 1029.

27. Brandenburg v. Ohio, 395 US 444, 23 L.ed. 2nd 430, 89 S.Ct. 1827, concurring.

28. The clear and present danger test has appeared recently in Nebraska Press Association v. Stuart, 427 US 539 (1976) and in Justice William H. Rehnquist's dissent in Central Hudson Gas and Electric Corp. v. Public Service Commission, 447 US 557 (1980).

29. Concurring opinion in Dennis v. US, 341 US 494, 95 L.ed. 1137, 71 S.Ct. 857.

30. Barenblatt v. US, 360 US 109, 3 L.ed 2d 1115, 79 S.Ct. 1081, dissenting. As Justice Stewart argued: "So long as Members of this Court view the First Amendment as no more than a set of values to be balanced against other values that Amendment will remain in grave jeopardy." Pittsburgh Press Co. v. Pittsburgh Commission on Human Relations, 413 US 376, 37 L.ed. 2d 669, 93 S.Ct. 2553, dissenting. See also Alexander Meiklejohn, "The First Amendment Is an Absolute," *Supreme Court Review*, 1961, pp. 245–266; Laurent Franz, "The First Amendment in the Balance," *Yale Law Journal* 71 (1962): 1424; Franz, "Is the First Amendment Law: A Reply to Mr. Mendelson," *California Law Review* 51 (1963): 729; but cf. in defense of the balance theory Wallace Mendelson, "On the Meaning of the First Amendment: Absolutes in the Balance," *California Law Review* 50 (1962): 821; Mendelson, "The First Amendment and the Judicial Process: A Reply to Mr. Franz," *Vanderbilt Law Review* 17 (1964): 479.

31. Bridges v. California, 314 US 252, 86 L.ed. 192, 62 S.Ct. 190. See also Douglas dissents in Kingsley Corp. v. Regents of University of New York, 360 US 684, 3 L.ed. 2d 1512, 79 S.Ct. 1362; Roth v. US, 354 US 476, 1 L.ed. 2d 1498, 77 S.Ct. 1304; concurrence in Superior Films v. Dept. of Education of State of Ohio, 346 US 587, 98 L.ed. 329, 74 S.Ct. 286; Black dissent in Carlson v. Landon, 342 US 524, 96 L.ed. 547, 72 S.Ct. 525.

32. Barenblatt v. US, 360 US 109, 3 L.ed. 2d 1115, 79 S.Ct. 1081. The difference in the words chosen by the framers for the First Amendment and for other clauses was noted by Justice Douglas as evidence in favor of an absolute interpretation: "The First Amendment is couched in absolute terms—freedom of speech shall not be abridged. Privacy, equally sacred to some, is protected by the Fourth Amendment only against unreasonable searches and seizures." Interference with speech is therefore barred, whether the constraint is reasonable or unreasonable. Dissent in Beauharnais v. Illinois, 343 US 250, 96 L.ed. 919, 72 S.Ct. 752.

33. For some of many examples of use of the balance metaphor by the Court majority, see Gertz v. Welch, 418 US 323, 41 L.ed. 789, 94 S.Ct. 2997 (1974); Lloyd Corp. v. Tenner, 407 US 551, 33 L.ed. 2d 131, 92 S.Ct. 2219, Barenblatt v. US, 360 US 109, 3 L.ed. 2d 1115, 79 S.Ct. 1081.

34. Frankfurter repeatedly protested this position. In a pained dissent against this doctrine in Kovacs v. Cooper, 336 US 77, 93 L.ed. 513, 69 S.Ct. 448 (1949), he summarized the history of the doctrine, tracing it first to Herndon v. Lowry, 301 US 242, 81 L.ed. 1066, 57 S.Ct. 732 (1937): "The

power of a state to abridge freedom of speech and of assembly is the exception rather than the rule." "The judgment of the legislature is not unfettered." The following year a footnote in US v. Carolene Products Co., 304
US 144, 82 L.ed. 1234, 58 S.Ct. 778 (1938), argued that the usual presumption
in favor of the constitutionality of legislation must be narrowed when it appears "to be within a specific prohibition of the Constitution, such as those
of the first ten amendments." This footnote was cited in Thornhill v. Alabama, 310 US 88, 84 L.ed. 1093, 60 S.Ct. 736 (1940); American Federation of
Labor v. Swing, 312 US 321, 85 L.ed. 855, 61 S.Ct. 568 (1941); Thomas v.
Collins, 323 US 516, 89 L.ed. 430, 65 S.Ct. 315 (1945). After restatement in
Schneider v. Irvington, 308 US 147, 84 L.ed. 155, 60 S.Ct. 146 (1939), the
special solicitude of the Court for the First Amendment received its classic
expression by Black in 1942 in Bridges v. California, 314 US 252, 86 L.ed.
192, 62 S.Ct. 190. The phrase "preferred position" thereafter appeared in a
number of decisions. See e.g. Marsh v. Alabama, 326 US 501, 90 L.ed. 265,
66 S.Ct. 276 (1946). Justice Frankfurter returned to the attack in 1951 in
concurrence in Dennis v. US, 341 US 494, 95 L.ed. 1137: "Some members of
the Court—and at times a majority . . . have suggested that our function in
reviewing statutes restricting freedom of expression differs sharply from our
normal duty in sitting in judgment on legislation . . . It has been weightily
reiterated that freedom of speech has a 'preferred position' among constitutional safeguards . . . We have . . . given constitutional support, over repeated protests, to uncritical libertarian generalities." The Court continued
to affirm that position without repeatedly using the phrase that incurred
their brother's attacks.

35. These are nine different but not disjoint rules. They overlap consid-
erably. Many of the examples given can be cited for more than one rule.
The rules differ in the principle that is applied, not always in the domain of
facts to which they refer.

36. Erznoznik v. Jacksonville, 422 US 205, 45 L.ed. 2d 125, 95 S.Ct. 2268.
In a case concerning *Lady Chatterley's Lover* the Court overturned a statute
that barred favorable portrayals of immoral behavior; the Court declined to
construe the statute as pertaining only to obscene favorable portrayals,
which would have rendered it constitutional. Kingsley International Pictures
Corp. v. Regents of University of State of New York, 360 US 684, 3 L.ed. 2d
1512, 79 S.Ct. 1362. The Court has also overturned local ordinances against
public use of "fighting words" such as name calling, unless the statutes in
question specify that the words are banned only when a threat to public
order. Cohen v. California, 403 US 15, 29 L.ed. 2d 284, 91 S.Ct. 1780 (1971);
Gooding v. Wilson, 405 US 518, 31 L.ed. 2d 408, 92 S.Ct. 1103 (1972);
Plummer v. Columbus, 414 US 2, 38 L.ed. 2d 3, 94 S.Ct. 17 (1974); Lewis v.
New Orleans, 415 US 130, 39 L.ed. 2d 214, 94 S.Ct. 970 (1974). On the contrary, in some Jehovah's Witness cases the Court ruled certain municipal ordinances invalid as applied to religious canvassing but did not overthrow
the ordinances in other applications. Murdock v. Pennsylvania, 319 US 105,

87 L.ed. 1292, 63 S.Ct. 870; Martin v. Struthers, 319 US 141, 87 L.ed. 1313, 63 S. Ct. 862; Follett v. McCormick, 321 US 573, 88 L.ed. 938, 64 S.Ct. 717. But in Hynes v. Mayor of Oradell, 425 US 610, 48 L.ed. 2d 243, 96 S.Ct. 1755, the Court on grounds of vagueness did throw out an ordinance requiring would-be canvassers to notify the police. A properly drafted ordinance, the Court said, would have been constitutional. The Court has more than one option open to it.

Historically, overthrowing rather than reinterpreting a badly drafted law is a novel posture for the Court. The more traditional and normal approach was taken in Fox v. Washington, 236 US 273, 59 L.ed. 573, 35 S.Ct. 383 (1915), when Fox, a nudist, published a denunciation of "prudes" who had called in the police to close a nudist colony. He was convicted for his remarks under a law prohibiting a publication encouraging or inciting the commission of any crime or "disrespect for law." Prohibition of direct incitement of crime is constitutional; prohibition of expressions of disdain for the legal system is not. As Justice Holmes wrote: "So far as statutes may fairly be construed in such a way as to avoid doubtful constitutional questions they should be so construed . . . It . . . is not likely that the statute will be construed to prevent publications merely because they tend to produce unfavorable opinions of law." But this decision has long since been overruled. Thornhill v. Alabama, 310 US 88, 84 L.ed. 1093, 60 S.Ct. 736 (1940); Winters v. New York, 333 US 507, 92 L.ed. 840, 68 S.Ct. 665 (1948). When First Amendment rights are at stake, the Court no longer feels bound to correct a badly written statute case by case.

37. A law defining Communists as lacking the moral character for admission to the bar was overturned. Konigsberg v. State Bar of California, 353 US 252, 1 L.ed. 2d 810, 77 S.Ct. 722. The immorality of each individual would have to be proved. Similarly a law requiring a non-Communist affidavit for tax exemption was overturned as reversing the proper presumption; the state has the burden of proving someone ineligible. Speiser v. Randall, 357 US 513, 2 L.ed. 2d 1460, 78 S.Ct. 1332.

38. Smith v. California, 361 US 147, 4 L.ed. 2d 205, 80 S.Ct. 215.

39. Island Trees Union Free School District v. Pico, *US Law Week* 50 (1982): 4831.

40. Harrison v. NAACP, 360 US 167, 176 (1959). In one case in 1971 a group had been distributing leaflets accusing a Mr. Keefe of "panic peddling," or blockbusting, and Keefe sought an injunction. An Illinois court granted a temporary injunction while considering the plea. The Supreme Court took the case at that stage instead of waiting to see what the state court would do because, it ruled, the injunction was an unconstitutional prior restraint. Organization for a Better Austin v. Keefe, 402 US 415, 29 L.ed. 2d 1, 91 S.Ct. 1575. In the Pentagon Papers case later in the same year the issue of urgency in dissolving an injunction also arose, and the Court acted with exceptional expedition. New York Times Co. v. US, 403 US 713, 29 L.ed. 2d 822, 91 S.Ct. 2140. The conventional position is that constitu-

tional questions should be avoided wherever possible. From this it follows that the Court "should not adjudicate the constitutionality of state enactments fairly open to interpretation until the state courts have been afforded a reasonable opportunity to pass on them." Douglas v. City of Jeannette, 319 US 157 (1943). But recently the Court, rather than postponing judicial review in First Amendment cases, has sometimes sped the process by which harassments of speech can be struck down.

41. See e.g. Smith v. California, 361 US 147, 151, 4 L.ed. 2d 205, 210, 80 S.Ct. 215: "This Court has intimated that stricter standards of permissible statutory vagueness may be applied to a statute having a potentially inhibiting effect on speech; a man may the less be required to act at his peril here because the free dissemination of ideas may be the loser." Winters v. New York, 333 US 507, 509, 510, 517, 518, 92 L.ed. 840, 846, 847, 850, 851, 68 S.Ct. 665. See also Bagett v. Bullitt, 377 US 360, 12 L.ed. 2d 377, 84 S.Ct. 1316; Dombrowski v. Pfister, 380 US 479, 14 L.ed. 2d 22, 85 S.Ct. 1116; dissent of Brennan, Stewart, and Marshall in Miller v. California, 413 US 15, 37 L.ed. 2d 419, 93 S.Ct. 2607; concurring opinion of Clark in Kingsley Corp. v. Regents of University of State of New York, 360 US 684, 3 L.ed. 2d 1512, 79 S.Ct. 1362; "The Void for Vagueness Doctrine in the Supreme Court," *University of Pennsylvania Law Review* 109 (1960–1961): 67.

42. In the first decades of this century the courts often used the Fourteenth Amendment to defend property rights against the denial of substantive due process arising from vagueness of legislation. Since President Franklin Roosevelt's shakeup of the Court, this defense of property rights has been rare. Today's priorities are expressed by the use of vagueness as a charge against laws infringing personal freedom.

43. See Coates v. City of Cincinnati, 402 US 611, 29 L.ed. 2d 214, 91 S.Ct. 1686; Smith v. California, 361 US 147, 4 L.ed. 2d 205, 80 S.Ct. 215; Staub v. Baxley, 355 US 313, 2 L.ed. 2d 302, 78 S.Ct. 277.

44. See "The First Amendment Overbreadth Doctrine," *Harvard Law Review* 83 (Feb. 1970): 844, 919: "The most energetic use of the overbreadth doctrine recently has been with respect to laws burdening subversive advocacy or affiliation. The Court has invalidated penal laws, loyalty oaths, and civil disabilities which single out persons who advocate certain political viewpoints or are affiliated with associations oriented to such advocacy." For penal laws, see Dombrowski v. Pfister, 380 US 479, 14 L.ed. 2nd 22, 85 S.Ct. 1116. While this decision did not overrule Dennis and similar cases that supported legislation singling out the Communist Party for restrictive action, it did throw out a Louisiana statute dealing with subversive organizations and Communist fronts as too broad. For loyalty oaths, see Keyishian v. Board of Regents, 385 US 589, 17 L.ed. 2d 629, 87 S.Ct. 675. For civil disabilities, see US v. Robel, 389 US 258, 19 L.ed. 2d 508, 88 S.Ct. 419. Even when the behavior of the defendant was of such a character that the state could constitutionally have made it punishable, laws that would also penalize protected behavior have been struck down as overbroad. Kunz v. N.Y. 340

US 290, 95 L.ed. 280, 71 S.Ct. 312. A regulation providing for seizure or suppression of obscene matter, for example, must be so drawn that no nonobscene publications are seized in the process, such as other issues of the same magazine. Marcus v. Search Warrant, 367 US 717, 6 L.ed. 2d 1127, 81 S.Ct. 1708; Lewis v. New Orleans, 415 US 130, 39 L.ed. 2d 214, 94 S.Ct. 970; Brennan, Stewart, Marshall dissent in Miller v. California, 413 US 15, 37 L.ed. 2d 419, 93 S.Ct. 2607; Paris Adult Theater 1 v. Slaton, 413 US 49, 37 L.ed. 2d 446, 93 S.Ct. 2628; Coates v. City of Cincinnati, 402 US 611, 29 L.ed. 2d 214, 91 S.Ct. 1686; Gooding v. Wilson, 405 US 518, 31 L.ed. 2nd 408, 92 S.Ct. 1103; Aptheker v. Secretary of State, 378 US 500, 12 L.ed. 2d 992, 84 S.Ct. 1659. Yet when the overbreadth of a law is relatively minor, the Court may choose to let the law stand, counting on proper interpretation. Broderick v. Oklahoma, 413 US 610; New York v. Ferber, *US Law Week* 50.50 (1982): 5077.

45. See e.g. Freedman v. Maryland, 380 US 51, 13 L.ed. 2d 649, 85 S.Ct. 734. Another set of procedural issues concerns a person's standing in court. To protest the constitutionality of an ordinance, law, or executive action, one must oneself be an injured party. The courts are not a forum for abstract discussion of public issues. This applies also in First Amendment cases, but here the Court has bent to give broader recognition of standing to defenders of free speech.

46. Shelton v. Tucker, 364 US 479, 5 L.ed. 2d 231, 81 S.Ct. 247 (1960).

47. Schneider v. Irvington, 308 US 147, 84 L.ed. 155, 60 S.Ct. 146. See also Smith v. California, 361 US 147, 151, 4 L.ed. 2d 205, 210, 80 S.Ct. 215 (1959).

48. In 1877 Justice Field in Ex parte Jackson, 96 US 727, 24 L.ed. 877, sustained the power of Congress to exclude certain types of matter from the mails by interpreting the congressional grant of a postal monopoly as not applying to matter that the post office declined to carry. In a 1949 case concerning the shipping of obscene phonograph records, it was argued that under a narrow construction of "obscenity" an object like a phonograph record was not covered because no visible or readable obscenity appeared. The Court did not buy this argument, but Frankfurter and Jackson dissented: "I cannot agree to any departure from the sound practice of narrowly construing statutes which by censorship restrict liberty of communication." US v. Alpers, 338 US 680, 94 L.ed. 545, 70 S.Ct. 352.

49. Bridges v. California, 314 US 252, 86 L.ed. 192, 62 S.Ct. 190.

50. Justice Rutledge in Thomas v. Collins, 323 US 516, 89 L.ed. 430, 65 S.Ct. 315.

51. Justice Murphy in Chaplinsky v. New Hampshire, 315 US 568, 86 L.ed. 1031, 62 S.Ct. 766 (1942).

52. Beauharnais v. Illinois, 343 US 250, 96 L.ed. 919, 72 S.Ct. 752.

53. Roth v. US, 354 US 476, 1 L.ed. 2d 1498, 77 S.Ct. 1304.

54. Kingsley International Pictures Corp. v. Regents of the University of the State of New York, 360 US 684, 3 L.ed. 2d 1512, 79 S.Ct. 1362.

55. The law in this area is in a state of chaos. Following Roth, a three-man plurality of the Court in Memoirs v. Massachusetts, 383 US 413, 16 L.ed. 2nd 1, 86 S.Ct. 975 (1966), extended First Amendment protection to material with even "the slightest redeeming social importance." See also Jacobellis v. Ohio, 378 US 184, 191, 12 L.ed. 2d 793, 800, 84 S.Ct. 1676 (1964). But faced with the loophole of a passage or two with a social message in otherwise hard-core pornography, the Court in 1973 retreated from the Roth criterion. Miller v. California, 413 US 15, 37 L.ed. 2d 419, 93 S.Ct. 2607. Declaring that it was finally going to solve the problem and sounding more like a legislature than a court, the majority in a 5–4 decision announced that it was laying down a new set of standards to define obscenity. The mishmash of criteria included the notion of local community standards, intended to get away from the impossible job of having the Court determine a national standard. As the criterion was applied to the work as a whole, it potentially denied First Amendment protection to works with a substantial but subordinate social message. Justice Douglas rejected the whole notion that obscenity defines a class of speech somehow excepted from the benefits of the First Amendment.

56. See FCC v. Pacifica Foundation, 438 US 726, 57 L.ed. 2d 1073, 98 S.Ct. 3026; Young v. American Mini-Theaters, 427 US 50; and New York v. Ferber, *US Law Week*, 50.50, (1982): 5077.

57. In 1915 the Supreme Court held movies to be just a business and to have no First Amendment rights. Mutual Film Corp. v. Industrial Commission, 236 US 230, 59 L.ed. 552, 35 S.Ct. 387. This view was overturned in 1952. Burstyn v. Wilson, 343 US 495, 96 L.ed. 1098, 72 S.Ct. 777. Schad v. Borough of Mount Ephraim, 101 S.Ct. 2176 (1981), upheld the application of the First Amendment to entertainment. Nonetheless, particular priority has been given to some cases dealing with repression of political dissidents, such as Lamont v. Postmaster General, 381 US 301, 14 L.ed. 2d 398, 85 S.Ct. 1493. Frankfurter, who more often than not voted on the antilibertarian side, made strong defenses of civil liberty whenever teachers and academic freedom were threatened. Scholarly inquiry, learning, and high culture constituted in his view the kernel of the speech that it was important to protect. Sweezy v. New Hampshire, 354 US 234, 1 L.ed. 2d 1311, 77 S.Ct. 1203.

For Alexander Meiklejohn's viewpoint, see his *Free Speech and Its Relation to Self-Government* (New York: Harper, 1948), reprinted in his *Political Freedom* (New York: Harper, 1960). It is particularly strongly expressed in New York Times v. Sullivan, 376 US 254, 11 L.ed. 2d 686, 84 S.Ct. 710 (1964), which concerned a libel suit by a public official whose actual behavior had apparently been misstated in an advertisement in the *Times.* Douglas expressed the consistent absolutist position that under the First Amendment no libel suit can be brought by a public official arising out of political accusations, no matter how inaccurate or scurrilous. The Court did not go that far but held that, to establish libel, a public figure, unlike an ordinary citizen in private libel suits, has to show malice. See Harry Kalvan, "The New York

Times Case: A Note on the Central Meaning of the First Amendment," *Supreme Court Review,* 1964, p. 191; Thomas Emerson, *Toward a General Theory of the First Amendment* (New York: Vintage, 1967). Recent cases have narrowed the protection given the press and the criteria of who is a public figure. Herbert v. Lando, 441 US 153 (1979); Wolston v. Readers Digest Association, Inc., 443 US 157 (1979); Hutchinson v. Proxmire, 443 US 111 (1979). But the basic law remains that of Sullivan.

58. Valentine v. Chrestensen, 316 US 51, 86 L.ed. 1262, 62 S.Ct. 920.

59. Cantwell v. Connecticut, 310 US 296, 84 L.ed. 213, 60 S.Ct. 900. See also Murdock v. Pennsylvania, 319 US 105, 87 L.ed. 1292, 63 S.Ct. 870.

60. Grossjean v. American Press Co., 297 US 233, 80 L.ed. 660, 56 S.Ct. 444.

61. See Smith v. California, 361 US 147, 4 L.ed. 2d 205, 80 S.Ct. 215. In sustaining the First Amendment rights of a union organizer, Justice Wiley Rutledge rejected the idea that "the First Amendment's safeguards are wholly inapplicable to business and economic activity." Thomas v. Collins, 323 US 516, 89 L.ed. 430, 63 S.Ct. 315. See also Justice Clark in Burstyn v. Wilson, 343 US 495, 96 L.ed. 1098, 72 S.Ct. 777 (1952): "It is urged that motion pictures do not fall within the First Amendment's aegis because their production, distribution, and exhibition is a large-scale business conducted for private profit. We cannot agree. That books, newspapers, and magazines are published and sold for profit does not prevent them being a form of expression whose liberty is safeguarded by the First Amendment. We fail to see why operation for profit should have any different effect in the case of motion pictures."

62. Breard v. Alexandria, 341 US 622, 95 L.ed. 1233, 71 S.Ct. 920.

63. Cammerano v. US, 358 US 498, 3 L.ed. 2d 462, 79 S.Ct. 524. By 1974 four judges in Lehman v. City of Shaker Heights, 418 US 298, 314 n. 6, 41 L.ed. 2nd 770, 94 S.Ct. 2714, and three in Pittsburgh Press Co. v. Pittsburgh Commission on Human Relations, 413 US 376, 393, 398, 401, 37 L.ed. 2nd 669, 93 S.Ct. 2553, had rejected the doctrine of exclusion of commercial speech from the First Amendment.

64. Bigelow v. Virginia, 421 US 820 n. 6, 44 L.ed. 2d 600, 95 S.Ct. 2222.

65. Virginia State Board of Pharmacy v. Virginia Citizens' Consumer Council, 425 US 748, 48 L.ed. 2d 346, 96 S.Ct. 1817.

66. First National Bank of Boston v. Bellotti, 435 US 765, 55 L.ed. 2nd 707, 98 S.Ct. 1407 (1978), decision by Justice Powell and concurrence by Chief Justice Burger.

67. *New York Times,* May 7, 1978.

68. The Supreme Court refused to accept this distinction. Zurcher v. Stanford Daily, 436 US 547, 56 L.ed. 2d 525, 98 S.Ct. 1970 (1978).

69. FCC v. Pacifica Foundation, 438 US 726, 57 L.ed. 2d 1073, 98 S.Ct. 3026 (1978); First National Bank of Boston v. Bellotti, 435 US 765, 55 L.ed. 2d 707, 98 S. Ct. 1407 (1978).

70. Giboney v. Empire Storage and Ice Co., 336 US 490, 93 L.ed. 834, 69

S.Ct. 684. See also Harry Kalvan, "Upon Rereading Mr. Justice Black," *UCLA Law Review* 14 (1967): 428.

71. Justice Rutledge in Thomas v. Collins, 323 US 516, 530, 89 L.ed. 430, 440, 65 S.Ct. 315 (1944).

5. Carriers and the First Amendment

1. The *Gazette* continued under five successive postmasters till 1741.

2. Daniel C. Roper, *The United States Post Office* (New York: Funk and Wagnalls, 1917), p. 4.

3. Roper, *The United States Post Office*, p. 6.

4. By a regulation of 1603 every postmaster was required to keep two horses on hand for government messages and to forward them within 15 minutes of receipt "at the rate of not less than 7 miles an hour in summer and 5 miles in winter." Roper, *The United States Post Office*, p. 9.

5. Richard K. Craille, ed., *Speeches of John C. Calhoun* (New York, 1864), p. 190. Speech delivered 1817.

6. Frank Luther Mott, *American Journalism* (New York: Macmillan, 1941), p. 178.

7. Wayne T. Fuller, *The American Mail: Enlarger of the Common Life* (Chicago: University of Chicago Press, 1972), p. 132; Anthony Smith, *Subsidies and the Press in Europe* (London: PEP, 1977).

8. The earliest practice in Europe had been to charge the receiver of mail. For reasons of ensuring payment, the practice shifted, but only with the introduction of the postage stamp in 1845 did the American postal system move almost fully toward prepayment.

9. Fuller, *The American Mail*, p. 114. A North Carolina congressman said in 1850, "The poisoned sentiments of the city, concentrated in their papers, with all the aggravation of such a moral and political cesspool, will invade the simple, pure, conservative atmosphere of the country, and meeting with no antidote in the rural press, will contaminate and ultimately destroy that purity of sentiment and of purpose, which is the only true conservatism." Along with the fear that urban immorality would penetrate the countryside through urban newspapers went the belief, expressed in a Senate report of 1832, that a monopoly of the city-controlled press, whose political atmosphere was "not always congenial to a spirit of independence," might undo democracy. "A concentration of political power in the hands of a few individuals is of all things, most to be dreaded in a republic. It is, of itself, an aristocracy more potent and dangerous than any other; and nothing will tend so effectually to prevent it as the sustaining of the newspaper establishments in the different towns and villages throughout the country." Urban legislators, on the other hand, demanded lower postage on interstate newspapers going more than 100 miles and tried in various other ways to eliminate the discriminations in favor of the rural press, arguing that the government's policy bred a sectional press, nurtured sectional opinions,

prevented citizens of one state from understanding those of another, and limited the spread of intelligence. But nothing could change the mind of Congress on this subject.

10. Congressional Committee on Post Roads, quoted in Mott, *American Journalism*, p. 194.

11. By 1970, when the American postal deficit was soaring toward a billion dollars, Congress finally began modifying this philosophy in the Postal Reorganization Act, which set as a goal the depolitization of the postal service and the introduction of business management methods. The service ceased being a government department and became the United States Postal Service. A Postal Rate Commission was empowered to fix rates at levels high enough that by 1985 each class of service would pay its own way. These reforms are still partly in question, and whether Congress will in the end sustain or retreat from them remains to be seen.

12. Reform legislation since the mid nineteenth century has increasingly denied the President the ability to use various carrots as sticks. Creation of a civil service restricting patronage was one such measure. So was the creation of independent bipartisan regulatory commissions, such as the FCC, beyond the President's control. No President may now fly in a plane at public expense to make a campaign speech, appoint a political agent to a dummy job, give a license to a political ally, or as Jefferson did, support a newspaper that favors him.

13. 1 *Statutes at Large* 32, Ch. 7; 5 *Statutes at Large* 736.

14. Roper, *The United States Post Office*, p. 10.

15. See Lindsey Rogers, *The Postal Power of Congress: A Study in Constitutional Expansion* (Baltimore: Johns Hopkins Press, 1916), pp. 103–114; Dorothy Ganfield Fowler, *Unmailable: Congress and the Post Office* (Athens: University of Georgia Press, 1977), pp. 26–33.

16. 12 How (US) 88, 13 L.ed. 905.

17. Art. 1, sec. 8, clause 1.

18. A liberal construction of federal fiscal powers was one of the basic tenets of the Federalist position which historically had profound influence on the judicial branch. In viewing the postal service as primarily a fiscal activity, the Court was building on British traditions, where public mails had been set up to help finance official correspondence. It remained policy in Great Britain for centuries to run the mails for revenue. In America the issue was debated from the start as to whether the main goal of the postal service should be revenue for the government or expansion of national communication. Only after about 1820 did the post office come out from under Treasury control and the expansive doctrine become the accepted one. Roper, *The United States Post Office*, pp. 8–9; Wesley E. Rich, *The History of the United States Post Office to the Year 1829* (Cambridge: Harvard University Press, 1924), pp. 162–166.

19. Ex parte Jackson, 96 US 727, 24 L.ed. 877. The petitioners argued:

"Congress has no power to prohibit the transmission of intelligence, public or private, through the mails; and any statute which distinguishes mailable from unmailable matter merely by the nature of the intelligence offered for transmission, is an unconstitutional enactment . . . Congress has the right and the exclusive right to carry the mail; but in its exercise . . . it cannot evade a duty which . . . is inseparable from the right . . . to receive, transport, and deliver all letters."

20. This view was shared by Congress, which in 1825 forbade the transport of letters, packages, or other mailable matter "between places, from one to the other of which mail is regularly conveyed." Congress did not attempt to restrict private distribution wherever regular mail service was not already in operation, but it did seek to set up classes of things that were or were not mailable within the government's service, wherever that service was offered. 5 *Statutes at Large* 736.

21. Calhoun's argument did not exclude definition of what was mailable by criteria such as size, shape, or weight.

22. Williams v. Wells Fargo and Co. Express, 177 F 352. The decision treats the right of Congress to establish a monopoly in delivering the mails, which was "at one time questioned," as settled but defines "mail" narrowly.

23. It could be argued that the erosion of freedom of the mails began as a Civil War measure, when the Postmaster General denied postal privileges to several papers for treasonable utterances. Rogers, *The Postal Powers of Congress*, p. 51; Fowler, *Unmailable*, pp. 42–52; Alfred M. Lee, *The Daily Newspaper in America* (New York: Macmillan, 1937), p. 305; Eberhard P. Deutsch, "Freedom of the Press and of the Mails," *Michigan Law Review* 36: 724–725.

24. Fowler, *Unmailable*, pp. 55–85.

25. Following similar logic, Calhoun in 1836 introduced a bill to ban from the mail in a state any material which that state had banned. The bill was rejected by the Senate.

26. Public Clearing House v. Coyne, 194 US 497, 48 L.ed. 1092, 24 S.Ct. 789 (1904). The history of fraud orders goes back to 1860 when Attorney General Black advised the Postmaster General that he could issue certain fraud orders even in the absence of congressional action. Deutsch, "Freedom of the Press," p. 726.

27. Boyd v. US, 116 US 616, 635, 29 L.ed. 746, 6 S.Ct. 524.

28. Public Clearing House v. Coyne, 194 US 497, 48 L. ed. 1092, 24 S.Ct. 789 (1904).

29. Lewis Publishing Co. v. Morgan, 220 US 288, 57 L.ed. 1190, 33 S.Ct. 867.

30. Milwaukee Social Democratic Publishing Co. v. Burleson, 255 US 407, 65 L.ed. 704, 41 S.Ct. 352 (1921).

31. Leach v. Carlile, 258 US 138, 66 L.ed. 511, 42 S.Ct. 227.

32. Pike v. Walker, 73 App. DC 289, 121 F. 2d 37 (1941).

33. Hannegan v. *Esquire*, 327 US 146, 90 L.ed. 586, 66 S.Ct. 456 (1945).

34. Blount v. Rizzi, 400 US 410, 27 L.ed. 2d 498, 91 S.Ct. 423. The quoted words are from Freedman v. Maryland, 380 US 51, 13 L.ed. 2d 649, 85 S.Ct. 734 (1965), on which Blount relies.

35. Lamont v. Postmaster General, 381 US 301, 14 L.ed. 2d 398, 85 S.Ct. 1493.

36. The post office argued that the filing of the suit was an affirmative reply to its inquiry and that, in light of its forwarding the material, the case was moot. But the Supreme Court, giving the First Amendment its preferred status under the law, departed from normal procedures and heard the case because a First Amendment issue was at stake.

37. Yet intolerance never dies. In 1981 journals from Cuba were held back by the postal service exactly as had been journals from China in the 1960s. Although the government backed down after a year and resumed delivery, the fact that the issue could arise after Lamont reveals how treacherous is a situation in which a carrier has not only a monopoly but also discretion about what it carries.

38. Blount v. Rizzi, 400 US 410, 416, 27 L.ed. 2d 498, 503. The internal quote is from Marcus v. Search Warrant, 367 US 717, 731, 6 L.ed. 2d 1127, 1135, 81 S.Ct. 1708 (1961).

39. But in Israel, Great Britain, and a few other countries one venturesome company known as Consortium Communications, Ltd., has gone into precisely this business and survived several lawsuits. The international telex rates have been so high that the company can pick up messages, send them in batch on voice-grade lines, and deliver them at the other end for less than telex. This sort of resale service is becoming common in the United States, but in Europe it is still controversial.

40. Estimates from 1890 place "the proportion of business telegrams as high as 95 percent and the speculative class accounting for nearly half of the total." David Seipp, *The Right to Privacy in American History,* Harvard Program on Information Resources, Publication No. P-78-30, July 1978, p. 105.

41. Art. 1, sec. 18, clause 3. Pensacola Telegraph Co. v. Western Union Telegraph Co., 96 US 1, 24 L.ed. 708 (1878). For radio, see Pulitzer Publishing Co. v. FCC, 94 F 2d 249, 68 App. D.C. 124 (1938); NBC v. US, 319 US 190, 87 L.ed. 1344, 63 S.Ct. 997 (1943).

42. The only civil liberty issue extensively debated regarding telegraphy was the right of the government to seize and read telegrams. This Fourth Amendment "search and seizure" issue began in the Civil War when the War Department seized the previous year's telegrams in several cities in a search for traitors. Congressional committees later subpoenaed telegrams for several investigations, including those on President Andrew Johnson's impeachment and the contested Hayes-Tilden election. There was much debate about whether a telegram deserved the same immunity as a sealed letter. Seipp, *The Right to Privacy,* pp. 30-41.

43. Robert L. Thompson, *Wiring a Continent* (Princeton: Princeton University Press, 1947), pp. 217, 221.

44. Alvin F. Harlow, *Old Wires and New Waves* (New York: D. Appleton-Century, 1936), p. 177.

45. Thompson, *Wiring a Continent*, p. 223.

46. Harlow, *Old Wires*, p. 178.

47. Thompson, *Wiring a Continent*, pp. 235–239.

48. Thompson, *Wiring a Continent*, p. 236.

49. Francis Williams, *Transmitting World News* (UNESCO: Paris, 1953), pp. 18–19.

50. Williams, *Transmitting World News*, p. 19.

51. Williams, *Transmitting World News*, p. 20.

52. Jeffrey Kieve, *The Electric Telegraph in the UK* (Newton Abbot: David and Charles, 1973), p. 71.

53. Williams, *Transmitting World News*, p. 20.

54. Kieve, *The Electric Telegraph*, p. 119.

55. Kieve, *The Electric Telegraph*, p. 218. In addition, 180 million duplicate words were sent at the second copy rate.

56. Kieve, *The Electric Telegraph*, p. 218.

57. Kieve, *The Electric Telegraph*, p. 224.

58. James M. Herring and Gerald C. Gross, *Telecommunications: Economics and Regulation* (New York: McGraw-Hill, 1936), p. 1.

59. Herring and Gross, *Telecommunications*. Figuring a pound at five dollars, American press rates were three times the British rates for the same period.

60. Lee, *The Daily Newspaper*, p. 509.

61. Primrose v. Western Union Telegraph Co., 154 US 1, 38 L.ed. 883, 14 S.Ct. 1098.

62. Hannibal and Saint Joseph Railroad Co. v Swift, 12 Wall 262, (1870); Philadelphia and Reading R.R. Co. v. Derby, 14 How. 468 (1852).

63. Despite the Fourteenth Amendment, the legality of regulating rates for businesses affected with the public interest was established in the Granger cases in 1877. Munn v. Illinois, 94 US 113, 24 L.ed. 77 (1877).

64. Daniel J. Czitrom, *Media and the American Mind from Morse to McLuhan* (Chapel Hill: University of North Carolina Press, 1982), pp. 26–28.

65. US Post Office Department, Annual Report of the Postmaster General of the United States (Washington, D.C.: Government Printing Office, 1872), p. 29.

66. People ex rel Western Union Telegraph Co. v. Public Service Commission, 230 NY 95, 129 NE 220, 12 ALR 960 (1920).

67. Attorney General v. Edison Telephone Co., 6 O.B. Div. 244.

68. Anon., *Central Law Journal* 10 (1880): 178; William G. Whipple, *Central Law Journal* 22 (1886): 33; W. W. Thornton, *American Law Register* 33 (1886): 327; Herbert H. Kellog, *Yale Law Journal* 4 (1894): 223; Chesapeake and Potomac Telephone Co. v. Baltimore and Ohio Telegraph Co., 66 Md. 339.

69. Richmond v. Southern Bell Telephone and Telegraph Co., 174 US 761, 43 L.ed. 1162, 19 S.Ct. 778.

70. Sullivan v. Kuykendall, S.Ct. of Kentucky, 1885, *American Law Review* 33: 448.

71. Globe Printing Co. v. Stahl, 23 Mo. App. 451 (1886). Bank of Yolo v. Sperry Flour Co., 141 Cal. 314 (1903); Young v. Seattle Transfer Co., 33 Wash. 225 (1903).

72. Primrose v. Western Union Telegraph Co., 154 US 1, 38 L.ed. 883, 14 S.Ct. 1098.

73. The cases on liability were reviewed by Brutus Clay in *Virginia Law Review* 337 (1914). He presents similar arguments to those in the telegraph case cited above for the courts generally declining to hold phone companies liable for all consequences of a failure of service.

74. Richard Gabel, "The Early Competitive Era in Telephone Communication, 1893–1920," *Law and Contemporary Problems* 34 (Spring 1969): 340–359.

75. One anti-Bell pamphlet, printed in millions of copies, was Paul Latzke, *A Fight with an Octopus* (Chicago: Telephony Publishing, 1906).

76. Sec. 614–52, Ohio General Code.

77. Celina and Mercer County Telephone Co. v. Union-Center Mutual Telephone Association, 102 Oh. St. 487, 133 NE 540, ALR 1145.

78. Munn v. Illinois, 94 US 113, 130, 24 L.ed. 77. See also Farmers' and Merchants' Co-operative Telephone Co. v. Boswell Telephone Co., 187 Ind. 371, 119 NE 513 (1918).

79. FCC v. RCA Communications, Inc., 346 US 86, 97 L.ed. 1470, 73 S.Ct. 998.

80. Hawaiian Telephone Co. v. FCC, 162 US App. DC 229, 498 F 2d 771.

81. See Grossjean v. American Press Co., 297 US 233, 80 L.ed. 660, 56 S.Ct. 444 (1936).

82. Pugh v. City and Suburban Telephone Co., 9 Bull 104: "If indecent or rude or improper language was permitted, evil and ill-disposed persons would have it in their power to use it as a medium of insult to others, and perchance by some accident, such as the crossing of wires, or by a species of induction, the same communication might be launched into the midst of some family circle under very mortifying circumstances." See also William H. Rockel, *American Law Register* 37 (1889): 73; Huffman v. March Mutual Telephone Co., 143 Iowa 590 (1909). The Communications Act of 1934 forbids telephoning any comment, request, suggestion, or proposal which is obscene, lewd, lascivious, filthy, or indecent.

6. Broadcasting and the First Amendment

1. Asa Briggs, *The History of Broadcasting in the United Kingdom*, vol. 1, *The Birth of Broadcasting* (London: Oxford University Press, 1961), pp. 47–48.

2. Briggs, *Birth of Broadcasting*, pp. 49, 53, 55. See also W. M. Daltan, *The Story of Radio* (London: Adam-Hilger, 1975), II, 50.

3. R. N. Vyvyan, *Wireless over Thirty Years* (London: George Routledge, 1933), pp. 204–205.

4. H.R. 13159, 65th Cong., 2nd Sess.; W. L. Rodgers, "The Effects of Cable and Radio Control on News and Commerce," *The Annals of the American Academy of Political and Social Science* 112 (Mar. 1924): 255.

5. Bruce Bliven, "How Radio Is Remaking Our World," *The Century Magazine* 108.2 (June 1924): 149.

6. Seymour N. Siegel, "Censorship in Radio," *Air Law Review* 7.1 (Jan. 1936): 21. See also Stanley High, "Radio Policy Disarms Its Critics," *Literary Digest* 118.6 (Aug. 11, 1934): 23: "The American public never would submit to the censorship which British broadcasting imposes. British radio speeches are subject to check and recheck."

7. M. D. Fagen, ed., *A History of Science and Engineering in the Bell System* (Bell Telephone Laboratories, 1976), I, 384.

8. Secretary of Commerce Herbert Hoover, opening address at Fourth National Radio Conference, in *Radio Control:* Hearings Before the Committee on Interstate Commerce, United States Senate, 69th Cong., 1st Sess., Jan.-Mar. 1926, p. 57.

9. *New York Times*, Nov. 9, 1925, p. 25.

10. Morris Ernst, "Radio Censorship and the 'Listening Millions,' " *The Nation*, Apr. 28, 1926, p. 473. Some authorities claimed that in 1927 there were already 300 more stations than optimal.

11. 37 Stat. 302, 1912; Hoover v. Intercity Radio, 286 Fed. 1003, Ct. of App. D.C. 1923.

12. Note, "Indirect Censorship of Radio Programs," *Yale Law Journal* 40 (1931): 971; Senate Hearings, 1926, pp. 217–220.

13. US v. Zenith Radio Corp., 12 F 2d 614 N.D. Ill. (1926).

14. Senate Hearings, 1926, p. 55.

15. Fagen, *Science and Engineering in the Bell System*, p. 406.

16. 68 Cong. Rec., 69th Cong., 2nd Sess., pp. 2572–2573a.

17. Senate Hearings, 1926, pp. 26–27.

18. Senate Hearings, 1926, pp. 55–57.

19. Hugo L. Black, "No Broadcasting by Utilities," *Public Utilities Fortnightly* 3.12 (June 13, 1929): 686–692.

20. Quoted in Bliven, "How Radio Is Remaking Our World," p. 154; *New York Times*, Apr. 13, 1924, p. 17.

21. *New York Times*, Dec. 28, 1924, sec. 8, p. 13.

22. *New York Times*, Sept. 17, 1925, p. 15; Nov. 9, 1925, p. 25. The resolution on censorship reads that since "the success of radio broadcasting is founded upon the maintenance of public good will" and "the public is quick to express its approval or disapproval of broadcast programs . . . any agency of program censorship other than public opinion is not necessary." Proceedings of the Fourth National Radio Conference, quoted in Senate Hearings, 1926, pp. 59–60.

23. Senate Hearings, 1926, pp. 50, 56–58. This notion of local franchising, now practiced for cable television, was not adopted for radio, presumably because radio waves do not stop at political boundaries.

24. Senate Hearings, 1926, pp. 25–26. The chairman of the Committee on Legislation was Judge Stephen B. Davis, Jr., solicitor of the Commerce Department and the official charged with issuance of station licenses.

25. House Committee on Merchant Marine and Fisheries Hearings on a Bill to Regulate Radio Communication, 69th Cong., 1st Sess., p. 39; 67 Cong. Rec. 5480, 69th Cong., 1st Sess.

26. Stephen B. Davis, Jr., Senate Hearings, 1926, p. 121.

27. Representative Davis, 67 Cong. Rec. 5484, 69th Cong., 1st Sess. Norman Baker, owner of a station in Muscatine, Iowa, said: "Who is going to decide what the public interest is? Is it that . . . those stations which give sort of a home-like program, or is it against the schools and colleges that have intellectual talks at various times?" Senate Hearings, 1926, p. 166.

28. *New York Times*, Apr. 29, 1926, p. 9.

29. Efforts of Senators Hiram Johnson and Robert La Follette to broadcast "were frustrated under peculiar circumstances" in which "the connections were so arranged that but few places heard the speakers." *New York Times*, May 7, 1926, p. 21.

30. 35 Op. Att. Gen. 132.

31. *New York Times*, July 21, 1926, p. 18.

32. 68 Cong. Rec. 3257, 69th Cong., 2nd Sess.

33. Public Law No. 632, 69th Cong., secs. 18, 29, 11, 13. This was a jab at AT&T.

34. Barnouw, *A History of Broadcasting*, I, 66–87.

35. Quoted in Carl Dreher, "Why Censorship of Programs is Unfortunate," *Radio Broadcast* 10 (1927): 278.

36. The *New York Times*, Sept. 8, 1924, editorialized that the law requiring newspapers to label advertising as such should be applied to radio since "broadcasting certainly is publishing." p. 14.

37. Proceedings of Third National Radio Conference, pp. 2–3; *New York Times*, Oct. 9, 1924, p. 25.

38. "Radio Inspectors Ought Not To Be Censors," *Radio Broadcasting* 5 (Aug. 1924): 299.

39. Bliven, "How Radio Is Remaking Our World," pp. 147–154. The *New York Times*, Apr. 9, 1924, p. 20, editorialized: "Yet mankind, so impotent in the presence of the new giant of electricity, must make the control of it, so far as it can be controlled, a function of Government. Private monopoly is unthinkable." "Private monopoly" in radio in 1924 was a code word for AT&T.

40. The *New York Times*, May 9, 1924, p. 22.

41. David Sarnoff, Grover Whalen, and Hudson Maxim, "The Freedom of the Air," *The Nation* 119 (July 23, 1924): 90. Maxim (p. 91) confessed to being puzzled: "I distrust the wisdom of allowing radio broadcasting to be

controlled by any private monopoly, but I also distrust the wisdom and the ability and the justice of federal control of radio."

42. Proceedings of Third National Radio Conference, pp. 13, 19; "Radio Censorship," *Literary Digest* 83 (Oct. 4, 1924): 28. This article opines that the public would protest at the idea of witnesses' words at a "vulgar trial" being transmitted through the air.

43. Barnouw, *A History of Broadcasting*, I, 140.

44. H. V. Kaltenborn, "On Being 'On the Air,'" *The Independent* 114 (May 23, 1925): 583, 584.

45. Quoted in Barnouw, *A History of Broadcasting*, I, 141.

46. Rep. Emmanuel Celler believed that sensible editing of material by stations should replace the widespread practice of cutting off speakers. Merlin Aylesworth, president of NBC, thought that freedom of the air did not include the right to "bore, insult, or outrage" the listening public. *New York Times*, Apr. 29, 1927, p. 14.

47. *New York Times*, May 18, 1926, p. 42; May 25, 1926, p. 23.

48. Senate Hearings, 1926, p. 133. The hearings were on bills for radio regulation filed by Sens. Dill and Howell. Rep. White had introduced a similar bill in the House which also incorporated some of the Fourth Conference's recommendations. The *New York Times*, Dec. 23, 1925, p. 18, worried editorially over this proposed concentration of regulatory power in Washington. But according to Dill's bill, "Nothing in this act shall be understood or construed to give the Secretary of Commerce the power of censorship over the radio communications or signals . . . except as herein specifically stated and declared, and no regulation or condition shall be promulgated . . . which shall interfere with the right of free speech and free entertainment by means of radio communications except as specifically stated." Senate Hearings, 1926, p. 133.

49. Ernst, "Radio Censorship," 475; Morris Ernst, "Who Shall Control the Air?" *The Nation* 122 (Apr. 21, 1926): 443. The Federation of Labor's representative, W. J. H. Strong, charged that the Commerce Department, by refusing to grant a wavelength to a labor group despite an explicit mandate to do so in the 1912 Radio Act, was "reserving for certain privileged persons here the cream of the whole business so far as commercial licenses are concerned," and the only way to combat the trend toward monopoly was to establish "multitudinous stations." To this end he proposed that no one corporation or individual be granted more than one license; that the public be given an opportunity to appeal to the Commerce Department and to be heard in license issuance, renewal, or revocation proceedings; that the antimonopoly provision of the bill be tightened so that connected companies might not dominate the airwaves; that the President not have the power to take over stations in wartime; that licenses not be assigned on the basis of priority of application but preference be given to nonprofit organizations; and that "at the time of expiration of a license the holder shall not be given any preference solely for that reason, on the theory that the holder of the

license must appear before the Department and justify on the basis of public benefit the continued renewal of the permit." Senate Hearings, 1926, pp. 206, 217. Morris Ernst, "Radio Censorship and the 'Listening Millions,'" pp. 473–474, 475.

50. Federal Radio Commission, Annual Report for Fiscal Year 1927, p. 6.

51. Note, "The Radio Act of 1927," *Columbia Law Review*, 1927, p. 732. See also *New York Times*, Sept. 18, 1927, sec. 10, p. 19. The *Times*, Apr. 29, 1927, p. 20, maintained that radio should be as open to free expression as the press but needed "high-minded direction."

52. Federal Radio Commission, Second Annual Report, pp. 153, 155, 159.

53. Federal Radio Commission, Second Annual Report, pp. 160–161.

54. Henry A. Bellows, "The Right to Use Radio," *Public Utilities Fortnightly* 3.13 (June 27, 1929): 773. The Third Annual Report of the Federal Radio Commission further elucidated the "public interest" standard in a way that implicated it in policies about content.

55. Edward C. Caldwell, "Censorship of Radio Programs," *Journal of Radio Law* 1.3 (Oct. 1931): 441, 470–472.

56. Commission on Communications: Hearings on S. 6, Senate Interstate Commerce Committee, 71st Cong., 1st Sess. (1929), pp. 1071–1073.

57. *New York Times*, Dec. 14, 1930, sec. 10, p. 17.

58. Paul Hutchinson, "Freedom of the Air," *The Christian Century* 48 (Mar. 25, 1931): 409.

59. Barnouw, *A History of Broadcasting*, I, 260.

60. Hutchinson, "Is the Air Already Monopolized?" *The Christian Century* 48 (Apr. 1, 1931): 442–443; Lauter and Friend, "Station Censorship," quoted in Harrison B. Summers, *Radio Censorship* (New York: H.W. Wilson, 1939), p. 158; Note, "Indirect Censorship of Radio Programs," p. 970.

61. Ansley v. Federal Radio Commission, 46 F 2d 600 D.C. Cir. (1930). The court noted that "much objectionable material had been broadcast" by the station. In Chicago Federation of Labor v. Federal Radio Commission, 41 F 2d 422, 423 D.C. Cir. (1930), when the labor organization sought to improve its power and hours of broadcasting after these had been reduced by the FCC, the court said that "the past record of station WCFL has not been above criticism."

62. KFKB Broadcasting Association v. Federal Radio Commission, 47 F 2d 670, 672 D.C. Cir. (1931).

63. Note, "Indirect Censorship of Radio Programs," p. 968. For preregulation or cautiously neutral positions, see Manuel Maxwell, Note, *Air Law Review* 2 (1931): 269; Howard W. Vesey, Note, *Journal of Radio Law* 131 (1931). Some liked the KFKB decision because it sanctioned the elimination of offensive advertising from radio.

64. Caldwell, "Censorship of Radio Programs," pp. 470, 473, distinguished between radio programs that were not technically "speech," such as entertainment and advertising, where the Radio Commission should have

broad discretion, and programs such as serious opinion, social, political, or economic commentary, and exposure of wrongs suffered by people, where any effort at censorship would contravene the First Amendment. He argued that the law provided ample means, such as libel suits, for dealing with the broadcast of illegal matter.

65. 62 F 2d 850 D.C. Cir. (1932); cert. den. 284 US 685 (1932); 288 US 599 (1933).

66. 62 F 2d 850 at 853 D.C. Cir. (1932). The reference to obstruction of justice alludes to the fact that Shuler had previously been jailed for contempt of court in connection with a radio broadcast commenting upon a pending case.

67. "Freedom for the Radio Pulpit," *The Christian Century* 49 (Jan. 27, 1932): 112, 113.

68. Note, *University of Pennsylvania Law Review* 81 (1933): 471, 471–72. See also Note, *Virginia Law Review* 19 (1933): 870; Note, "Refusal to Renew Broadcasting License as Restricting Freedom of Speech," *Duke Bar Association Journal* 1 (1933): 49; Note, *Air Law Review* 4 (1933): 96.

69. Note, "The Power of the Federal Radio Commission to Regulate or Censor Radio Broadcasts," *George Washington Law Review* 1 (1933): 384. The limitation of speech in the Trinity Methodist Church case was analogized to the sustaining of the Espionage Act and criminal syndicalism statutes. Note, 46 *Harvard Law Review* 46 (1933): 987.

70. Editorial, *Lexington Herald*, Nov. 22, 1933, quoted in Byron Pumphrey, "Censorship of Radio Programs and Freedom of Speech," *Kentucky Law Journal* 22 (1934): 640.

71. Note, "Radio Censorship and the Federal Communications Commission," *Columbia Law Review* 39 (1939): 455; Mitchell Dawson, "Censorship on the Air," *American Mercury*, Mar. 1934, pp. 257–268, reprinted in Summers, *Radio Censorship*. See also Minna Kassner, *Radio Is Censored* (New York: American Civil Liberties Union, 1936), p. 16.

72. Kassner, *Radio Is Censored*, pp. 13, 18.

73. 78 Cong. Rec. 2646–2648; Cong. Rec. 862–864, 73rd Cong., 2d Sess.; speech of Rep. McGugin, 78 Cong. Rec. 10327, 73rd Cong., 2d Sess.

74. 78 Cong. Rec. 10504, 73rd Cong., 2d Sess.

75. *New York Times*, Mar. 1, 1934; Apr. 21, 1934, p. 13.

76. FCC v. Pottsville Broadcasting Co., 309 US 134, 84 L.ed. 656, 60 S.Ct. 437. Indeed, the 1934 Communications Act stated its purpose to be "control of the United States over all the channels of interstate and foreign radio transmission."

77. 319 US 190, 87 L.ed. 1344. Justice Murphy dissented, largely on First Amendment grounds. The FCC imposed its will on the networks via rules on the stations it licensed because the 1934 Communications Act gave Congress no power to regulate networks directly. More recently, however, in rules about financial interests and syndication, the FCC has expanded its powers to regulate networks directly.

78. Red Lion Broadcasting Co. v. FCC, 395 US 367, 23 L.ed. 2d 371, 89 S.Ct. 1794 (1969). See also Fred W. Friendly, *The Good Guys, the Bad Guys and the First Amendment* (New York: Random House, 1975); Benno C. Schmidt, Jr., *Freedom of the Press vs. Public Access* (New York: Praeger, 1976); Henry Geller, *The Fairness Doctrine in Broadcasting*, RAND Report R–1412–77 (Santa Monica: The RAND Corp., 1973).

79. Those words in the decision are quoted from Senate Report No. 562, 86th Cong., First Sess., 8–9 (1959).

80. Jerome Barron, *Freedom of the Press for Whom?* (Bloomington: Indiana University Press, 1973); D. M. Gilmor and J. A. Barron, *Mass Communications Law* (Saint Paul: West Publishing, 1969); J. A. Barron, "An Emerging First Amendment Right of Access to the Media?" *George Washington Law Review* 37 (1969): 487. For the role of the citizen groups, see Marcus Cohn, "Who Really Controls Television?" *University of Miami Law Review* 29.3 (1975): 482–486.

81. Banzhaf v. FCC, 405 F 2d 1082 D.C. Cir. (1968).

82. Fred Friendly asserts that as soon as the law was passed and both cigarette advertising and the warnings against it went off the air, the sale of cigarettes increased, indicating the perverse results of some regulations. Friendly, *The Good Guys*, p. 110.

83. CBS, Inc., v. Democratic National Committee, 412 US 94, 36 L.ed. 2d 772, 93 S.Ct. 2080.

84. Note the remarkable statement that Congress, not the Constitution, gave broadcast editors their rights.

85. Miami Herald Publishing Co. v. Tornillo, 418 US 241, 41 L.ed. 2d 730, 94 S.Ct. 2831. Barron was Tornillo's lawyer.

86. Henri Blin et al, *Droit de la presse* (Paris: Librairies techniques, 1979); Jean Marie Auby and Robert Ducos-Ader, *Droit de l'information* (Paris: Dalloz, 1976).

87. Most particularly in the Democratic National Committee and the Family Viewing Hour cases.

88. Borrow v. FCC, 285 F 2d 666, 109 US App DC 224.

89. Lafayette Radio Electronics Corp. v. US, 345 F 2d 278 (1965).

90. New Jersey State Lottery Commission v. US, 420 US 371, 43 L.ed. 2d 260, 95 S.Ct. 941.

91. FCC v. Pacifica Foundation, 438 US 726, 57 L.ed. 2d 1073, 98 S.Ct. 3026 (1978).

92. Writers Guild of America, West, Inc., v. FCC, 423 F. Supp. 1064 (1976). The Ninth Circuit on appeal reversed the decision on the matter of primary jurisdiction and sent the case back to the FCC for consideration. Writers Guild of America, West, Inc., v. ABC, 609 F 2d 355, 9th Cir. (1974).

93. *Proceedings of the Third National Radio Conference* (Washington, D.C.: Government Printing Office, 1924), pp. 2–4. For other Hoover attacks on radio monopoly addressed at AT&T, see Paul Hutchinson, "Can the Air Be

Kept Free?" *Christian Century* 48 (1931): 549–550. Hutchinson notes that the networks, which by 1931 had become large organizations, never particularly liked this Hoover quotation because it could be interpreted as applying to them. President Coolidge also promised the conference delegates that the government would prevent a radio monopoly. *New York Times,* Oct. 9, 1924, p. 25.

94. 67 Cong. Rec. 5484 (1926).

95. Radio Act of 1927, Pub. L. No. 632, sec. 17 (1927). In debates at the time the statement was frequently made that broadcasting was a public utility but not a common carrier. As long as there was a scarcity of broadcast channels, and thus as long as licensees had a privileged position in the flow of public communications, it was hard to deny that in some respects broadcasting was a public utility, affected by the public interest. But to call it a common carrier was to support AT&T's bid for monopoly control of transmission facilities and thereby to oppose the new small broadcasting entrepreneurs. Cross-examination of W. D. Terrell of the Department of Commerce by Sen. Burton Wheeler and others at Senate hearings in 1929, illustrates the confused effort to distinguish between radio as a public utility and as a common carrier:

> *"Sen. Wheeler.* That is my view of it. These broadcasting stations are public utilities . . .
> *"Mr. Terrell.* There is quite a difference between the telephone, the telegraph and the radio. When you are talking by radio you are talking into the ears of everyone in the entire country . . . It is a common carrier in one way and not in another.
> *"Mr. Green.* Do you not mean it is a public utility and not a common carrier?"

Commission on Communications: Hearings on S.6 before the Senate Committee on Interstate Commerce, 71st Cong., 1st and 2d Sess. (1929), 1072–1073. See also Report of Committee No. 8 on legislation of the Fourth National Radio Conference, Senate Hearings, 1926, p. 25.

96. National Radio Coordinating Committee, letter to senators, Dec. 2, 1926, from Calvin Coolidge Papers, reel 80, case 136. See also Gavan Duffy, "Interests, Public and Private: The Development of Commercial Control of the Means of Transmission," paper, International Communications Association, Boston, Mass., May 3, 1982.

97. Leo Herzel, " 'Public Interest' and the Market in Color Television Regulation," *The University of Chicago Law Review* 18 (1951): 802–809. The classic treatment of economic allocation of spectrum through a market is R. H. Coase, "The Federal Communications Commission," *Journal of Law and Economics* 2 (1959): 1–40. Herzel's paper provoked a sharp response from

Dallas Smythe, "Facing Facts about the Broadcast Business," *University of Chicago Law Review* 20 (1952): 96; see also the rejoinder by Herzel, *University of Chicago Law Review* 20 (1952): 106. Smythe's left-wing criticism of the market had its mirror image on the right. Radio engineers have generally opposed the idea of a spectrum market out of a conviction that technicians can manage the allotments better than can a blind process.

98. Coase, "The Federal Communications Commission," p. 17.

99. Harvey J. Levin, *Fact and Fantasy in Television Regulation* (New York: Russell Sage Foundation, 1980), pp. 112, 118–119.

100. Red Lion Broadcasting Co. v. FCC, 395 US 367, 398, 23 L.ed. 2d 371, 89 S.Ct. 1794 (1969).

101. 47 U.S.C. 312 (a) (7); 47 C.F.R. Sec. 73. 1940 (1980). The Supreme Court in 1981 sustained this provision in CBS, Inc. v. FCC, 101 S.Ct. 2813 (1981).

102. Jora R. Minasian, "Property Rights in Radiation: An Alternative Approach to Radio Frequency Allocation," *Journal of Law and Economics* 18 (1975): 221; A. De Vany et al., "Electromagnetic Spectrum Management: Alternatives and Experiments," Appendix G to Staff Paper 7, President's Task Force on Communications Policy, 1967; Harvey Levin, *The Invisible Resource* (Baltimore: Johns Hopkins Press, 1971); Charles L. Jackson, "Technology for Spectrum Markets," Ph.D. thesis, MIT, 1976; Glen O. Robinson, dissenting opinion, Cowles Broadcasting, FCC 76–642; Carson E. Agnew, Richard G. Gould, Donald A. Dunn, and Robert D. Stibolt, "Economic Techniques for Spectrum Management," Report to NTIA by Mathtech, Inc., 1979; Douglas Rohall, "The Market Alternative in Radio Spectrum Allocation: Serving the Public Interest, Convenience and Necessity," Bachelor's thesis, MIT, 1982.

103. There is precedent for requiring beneficiaries of public grants to operate as common carriers. Under the Post Roads Act of 1866 the federal government gave rights to use routes over public lands only to telegraph companies which would operate as common carriers.

104. Jackson, "Technology for Spectrum Markets".

105. William P. McLauchlan and Richard M. Westerberg, "Allocating Broadcast Spectrum," *Telecommunications Policy* 6.2 (June 1982): 111–122.

106. FCC, *Federal Register* 46 (Feb. 24, 1981): 13888. See also McLauchlan and Westerberg, "Allocating Broadcast Spectrum," p. 118.

7. Cable Television and the End of Scarcity

1. The FCC also considered making directional assignments for FM stations to allow more to be licensed.

2. Rep. Lionel Van Deerlin made such a proposal. Jackson, *Technology for Spectrum Markets*, pp. 17, 24. Jackson recently proposed reallocation of part of the industrial band from 902 to 928 MHz to provide 1000 radio

broadcasting channels. Testimony to Senate Committee on Commerce, Science, and Technology, Sept. 28, 1982.

3. A pixel is a spot of color or black and white on a television screen. Under the American television standard, there are 525 lines of them. Each pixel is scanned 30 times a second under a scheme of scanning alternate ones, so the picture is scanned 60 times a second.

4. *Intermedia* 9.4 (July 1981).

5. Steven Rivkin, *A New Guide to Federal Cable Television Regulations* (Cambridge: MIT Press, 1978).

6. Conley Electric Corp. v. FCC, 394 F 2d 620. The Court denied the constitutional protest against the rule by the CATV system from Liberal, Kansas, first because the plea came too late, not having been raised in the FCC hearings, even though disregard of the normal rule on this matter is one of the special procedural features that marks the preferred status of the First Amendment. The Court went on, however, to reject the plea itself as without merit. See also Black Hills Video Corp. v. FCC, 399 F 2d 65; Titusville Cable TV, Inc., v. US, 404 F 2d 1187; Great Falls Community TV Cable Co. v. FCC, 416 F 2d 238. On the related matter of exclusivity, see Home Box Office, Inc., v. FCC, 587 F 2d 1248 (1978).

7. Rivkin, *A New Guide*, pp. 60–68.

8. Under later rules no charge was allowed for public access programs of less than 5 minutes. Rivkin, *A New Guide*, p. 82.

9. Rivkin, *A New Guide*, p. 83; 49 FCC 2d 1030 (1974).

10. The argument by cablecasters that Congress had not authorized regulation of them because they were neither common carriers as defined by one chapter of the act nor broadcasters as defined in another was rejected by the Court. US v. Southwestern Cable Co., 329 US 157, 20 L.ed. 2d 1001, 88 S.Ct. 1994.

11. Memorandum Opinion and Order in Docket 18397, 23 FCC 2d 825, *Federal Register* 35 (1970) 1091, quoted in Jackson, Shooshan, and Wilson, *Newspapers and Videotex*, pp. 28–29. See also First Report and Order in Docket 18397, 20 FCC 2d 201, 34 Fed. Reg. 17651 (1969).

12. Docket 19988.

13. Docket 19995, 79 FCC 2d 663.

14. Docket 20487.

15. Docket 20508.

16. NBC v. US, 319 US 190, 87 L.ed. 1344.

17. FCC v. Pottsville Broadcasting Co., 309 US 134, 84 L.ed. 656, 60 S.Ct. 437 (1940). The House Committee report included an incautious phrase to the effect that the purpose of the bill was to create a commission "armed with adequate statutory powers to regulate all forms of communication." HR Rep. No. 1850, 73rd Cong, 2d Sess., p. 3, quoted in US v. Southwestern Cable Co., 392 US 157, 20 L.ed. 2d 1001, 88 S.Ct. 1994. The Senate report was more careful, saying that the purpose of the bill was "to create a com-

munications commission with regulatory power over all forms of electrical communication." S. Rept. 781, 73rd Cong., 2d Sess., p. 1, quoted in Allen B. Dumont Laboratories v. Carroll, 184 F 2d 153, C.A. 3rd Circ (1950), which puts in the missing word "electrical" and softens the phrasing of the grant of power. The House version was clearly not within the constitutional powers of Congress, nor is there any evidence that such was ever intended; the statement was nothing more than a slip of the pen.

18. Harlan dissenting in 1906 in Patterson v. Colorado, 205 US 454, 51 L.ed. 879, 27 S.Ct. 556.

19. US v. Southwestern Cable Co., 392 US 157, 20 L.ed. 2d 1001, 88 S.Ct. 1994 (1968).

20. Paul J. Berman argues that the indivisibility view was implicitly rejected in 1974 in Teleprompter v. CBS, Inc., 415 US 394, 94 S.Ct. 1129, a case in which the issue was whether cable repeating was on the performing end of the stream and thus subject to copyright or on the receiving end and thus not subject. The Court held that while CATV systems could operate as program originators, the same system in other roles could function as a passive receiver. Berman, *CATV Leased-Access Channels and the Federal Communications Commission: The Intractable Jurisdictional Question*, Working Paper 75–6, Harvard University, Program on Information Technologies and Public Policy (1975).

21. White Smith v. Apollo, 209 US 1 (1908); Teleprompter Corp. v. CBS, Inc., 415 US 394, 94 S.Ct. 1129 (1974).

22. Rivkin, *A New Guide*, p. 72.

23. Black Hills Video Corp. v. FCC, 399 F 2d 65.

24. A staff report of the House Subcommittee on Communications concluded that CATV should not be regulated merely if broadcasting needed protection but only if the interest of the public as a whole would suffer in the absence of regulation. In related hearings Professor Anthony Oettinger pled with Congress to reconsider the limited mandate of the 1934 Communications Act so as to "restore primacy to the constitutional principles of the First Amendment, wherever the economic chips may fall." Rivkin, *A New Guide*, pp. 11, 14, 53. The District of Columbia Court of Appeals dealt with the same issue in the form of alleged harm by CATV to movies: "Even if the rules did reduce the number of films produced, that effect would not rise to the level of a First Amendment violation." Home Box Office, Inc., v. FCC, 567 F 2d 9 (1977).

25. US v. Southwestern Cable Co., 392 US 157, 20 L.ed. 2d 1001, 88 S.Ct. 1994 (1968). In a 1959 Report and Order re CATV and Repeater Services, Docket 12443, the FCC had reached a very different and more persuasive conclusion when it rejected four arguments for assuming jurisdiction over CATV. The first was that the systems were involved in broadcasting, a conclusion for which the FCC found no basis in the 1934 Communications Act. The second was that the systems were involved in common carriage, which the FCC had earlier rejected in Frontier Broadcasting Co. v. Collier,

24 FCC 451 (1958). The third was that the FCC had plenary power to regulate interstate communications; the FCC denied its power to regulate "any and all enterprises which happen to be connected with one of the many aspects of communications." The fourth was CATV systems' use of microwave relays. This rationale, if accepted, would restrict the contents of common carrier traffic and extend FCC jurisdiction indirectly to something it had no authority to regulate directly. Berman, *CATV Leased Access Channels,* pp. 9–10.

26. US v. Midwest Video Corp., 406 US 649. The issue was whether the FCC had authority to require cable systems that retransmitted broadcasts to originate some of their own programing as well. Without a concurring opinion by Chief Justice Warren Burger that sustained the FCC's authority but contained the quoted warning, the vote in the Supreme Court would have been a 4–4 tie. Burger called on Congress for a "comprehensive re-examination" of the Communications Act. Douglas dissented, rejecting totally the notion that cable carriers could be ordered by the FCC to become program producers. Another exploration of the limits of the FCC's authority over cable was National Association of Regulatory Commissioners v. FCC, 533 F 2d 601, in which the issue was whether, in the absence of explicit statutory authority, the FCC could exclude state governments from making access rules for CATV systems more extensive than but compatible with the federal rules. The appeals court rejected the FCC's attempt to claim exclusive jurisdiction over access rules in the absence of statutory language to that effect.

27. Home Box Office v. FCC, 567 F 2d 9, 185 U.S. App. D.C. 142 (1977). See also Miami Herald Publishing Co. v. Tornillo, 418 US 241, 94 S.Ct. 2831, 41 L.ed. 2d 730 (1974).

28. Midwest Video Corp. v. FCC, 571 F 2d 1025, 1052 (1978).

29. Midwest Video Corp. v. FCC (1978), 1054–1056, argues that the relationship of cablecasting to broadcasting "would appear far too tenuous and uncertain to warrant a cavalier overriding of First Amendment rights present in cablecasting."

30. Midwest Video v. FCC (1978) 1054.

31. The word "cable" is used in the discussion that follows as shorthand for any kind of broadband carrier that sends signals through an enclosed conduit which does not allow the signals to leak into the open where they will interfere with other signals. Whether coaxial cable, optical fibers, or new techniques are used, and whether these are operated by phone companies or by their competitors, will change over time as technology progresses. This discussion applies to all of them.

Optical fibers would seem to be the candidate medium for broadband networks except for the as yet unsolved problem of their untapability. Unlike a coaxial cable to which a feed to a customer can be attached without cutting the cable, an optical fiber can now only be entered by physically cutting it and interrupting service to others on the trunk.

32. Richard M. Neustadt, Gregg P. Skall, and Michael Hammer, "The Regulation of Electronic Publishing," *Federal Communications Law Journal* 33.3 (Summer 1981): 392. In 49 cities in the United States there are competing electric power companies having parallel sets of wires down the street. Walter J. Primeaux, Jr., "A Re-examination of the Monopoly Market Structure for Electric Utilities," in Almarin Phillips, ed., *Promoting Competition in Regulated Markets* (Washington, D.C.: Brookings, 1975), pp. 175–200. A handful of similar cases exist in cable television.

33. *On the Cable* (New York: McGraw-Hill, 1971).

34. Cabinet Committee on Cable Communications, *Cable—Report to the President* (Washington, D.C.: Government Printing Office, 1974). Clay T. Whitehead was the Director of the Office of Communications Policy in the Executive Office of the President and chairman of the Cabinet Committee. For other statements of the case for a common carrier system, see J. H. Barton, D. A. Dunn, E. B. Parker, and J. N. Rosse, *Nondiscriminatory Access to Cable Television Channels*, Program in Information Technology and Telecommunications, Report No. 2, Stanford University, 1973; Mark Nadel, "A Unified Theory of the First Amendment: Divorcing the Medium from the Message," *Fordham Urban Law Journal*, February 1983; Owen, *Economics and Freedom*, p. 136. See also Robert A. Kreiss, "Deregulation of Cable Television and the Problem of Access under the First Amendment," *Southern California Law Review* 54 (1981): 1001.

35. A staff report by a committee of the House of Representatives suggested a similar approach, proposing that after ten years cable operators be barred from doing programing and only allowed to provide channel facilities. Subcommittee on Communications of the House Committee on Interstate and Foreign Commerce, 94th Cong., 2d Sess., *Cable Television Promise Versus Regulatory Performance* (Washington, D.C.: Government Printing Office, 1976).

36. "We reaffirm our view that cable systems are neither broadcasters nor common carriers within the meaning of the Communications Act. Rather cable is a hybrid that requires identification and regulation as a separate force in communications." 36 FCC 2d 143, para. 191, quoted in Rivkin, *A New Guide*, p. 11.

37. First Report and Order on Program Origination, 20 FCC 2nd 202–203 (1969).

38. Today the general issue of how a communications company that has a monopoly in one part of the business can be allowed to engage also in competitive parts of the business without discriminating in its own favor is discussed largely in connection with telephony rather than cable. Proposals are being considered to separate in one way or another the more or less monopolistic plant that provides basic telephone transmission services from the competitive company or subsidiary that offers substantive services with a content. This issue, which is already a hot one with telephony, is likely to become hotter still with cable.

39. These provisions were clauses in the Senate version of the Communications Act rewrite, S.898, which were removed in amendments by Sen. Barry Goldwater. Negotiations are taking place at the moment of publication between the representatives of cities and the National Cable Television Association to seek agreement on a bill that both would back.

40. Kreiss, "Deregulation of Cable Television." The right of private newspaper owners to refuse unwanted ads has been sustained in Chicago Joint Board, Amalgamated Clothing Workers v. Chicago Tribune Co., 435 F 2d 470, 7 Cir. (1970), and in PMP Associates v. Boston Globe Newspaper Co., 321 NE 2d 915; but even that right holds only if not used for monopolization. The right of the state to create such exclusive rights is another matter.

41. Rivkin, *A New Guide*, p. 15, quoting from Whitehead Report. A study of arrangements between newspaper publishers and cable systems described 54 cases. Kathleen Criner and Raymond B. Gallagher, *Newspaper-Cable TV Services* (ANPA, March 1982).

42. Deborah Estrin, *Data Communications via Cable Television Networks: Technical and Policy Considerations,* Laboratory for Computer Science, MIT, Technical Report 273, May 1982; Marvin Sirbu and Deborah Estrin, *Alternative Technologies for Regional Communications: Technical, Economic, and Regulatory Issues: Data Communication over Cable,* working paper, Center For Policy Alternatives, MIT, May 1982. For such minimal two-way services as pay-for-view and alarm monitoring only the very small amount of upstream bandwidth needed to poll the subscribers is required. But for data communication, packet or other sophisticated multiplexing is required, and the system must be configured with more upstream bandwidth, a more sophisticated head-end computer, and other access points for users' computers.

43. In fact three not two major changes lead to the future capabilities of phone networks: broadband transmission via such carriers as optical fibers, digital transmission and switching, and common control switching. The common control system is a separate network, presumably a packet network, linking the same computers that serve as the digital switches of the regular phone network. The common control network is used to set up the calls on the regular customer network. One significance of this network is that it can be programed to do switching operations efficiently that would burden the phone system if they had to be done using the channels designed for voice. This is how such complicated operations can be set up as call forwarding to travelers wherever they may be in the country.

44. Cable is a local medium, so its costs are for local service. If there were 5000 cable systems in the country, to put something on all of them at $20 each would cost $100,000, plus the cost of the satellite transponder. This corresponds to the cost of time on a national television network. Cable is indeed more expensive than broadcasting over the air for sending a mass message to the whole population at once. What cable is good for is reaching small groups at reasonable cost.

45. Stanley M. Besen and Leland L. Johnson, *An Economic Analysis of Leased Channel Access for Cable Television* (Santa Monica: RAND Corp., forthcoming). This study, which signals the rising interest in the leasing problem, notes that without control on rates, compulsory leased access would not lead to diversity, but as it fails to note alternative ways of dealing with the rate problem, it reaches defeatist conclusions about leasing.

Two lower court decisions control for the moment who has authority to set rates for cable subscribers under the 1934 Communications Act and inferentially therefore the authority over rates for channel leases. Brookhaven Cable TV, Inc., v. Kelly, 573 F 2d 763, 2nd Cir. (1978), upheld the FCC's preemption of regulation of rates for pay programing, such as movies and sports, which is ancillary to broadcasting. National Association of Regulatory Utility Commissioners v. FCC (NARUC II), 533 F 2d 601, D.C. Circ. (1976), however, denied the FCC control over intrastate rates for nonvideo signals since these are covered by the carrier portion of the 1934 act, not the broadcasting portion. These legislative nonconstitutional interpretations are subject to change by congressional action.

46. J. H. Barton et al., *Nondiscriminatory Access to Cable TV Channels*, p. 6.

47. Cf. Bruce Owen, Jack Beebe, and Willard Manning, *Television Economics* (Lexington, Mass.: Lexington Books, 1974).

48. Allowing pay cable has been a controversial issue. The early opposition to it came largely from broadcasters, who saw its enormous revenue-producing capabilities as eclipsing their own entertainment business. In 1964 in California the cinema and television industries promoted a referendum outlawing any kind of pay television in the state. The proposition won, but the Supreme Court of California overturned it on First Amendment grounds. "It is no matter," the Court said, "that the dissemination takes place under commercial auspices." To make commercialism a basis for disallowing pay television "is comparable to asserting that no prohibition of expression would exist in the case of newspapers or motion pictures if a statute were adopted requiring their distribution or showing without charge." Weaver v. Jordan, 64 Cal 2d 235, 49 Cal. Rptr. 537, 411 P 2d 289 (1966). The National Association of Theatre Owners also challenged a decision by the FCC to allow pay television, charging that it violated the First Amendment both by depriving the poor of some communications and by surrounding the permission with antisiphoning rules. The appeals court rejected this challenge, citing the antisiphoning restrictions as the basis for concluding that the poor were not being deprived and finding the limiting regulations reasonable. National Association of Theatre Owners v. FCC, 420 F 2d 194, 136 US App DC 352 (1969).

8. Electronic Publishing

1. Charles M. Goldstein, "Optical Disk Technology and Information," *Science* 215.4534 (Feb. 12, 1982): 863–866.

2. John D. Chisholm and Terry Eastham, "Worldwide Network Reaps Enormous Savings," *Telecommunications*, Dec. 1978, pp. 53–56.

3. Carl F. J. Overhage and R. Joyce Harmon, eds., *Project Intrex: Report of a Planning Conference* (Cambridge: MIT Press, 1965).

4. The most successful device to date is the Kurzweil machine, which learns to read most fonts and currently provides input at about $1 a page.

5. Donald W. King, "Electronic Alternatives to Paper-based Publishing in Science and Technology," in Philip Hills, ed., *The Future of the Printed Word* (Westport, Conn.: Greenwood Press, 1980), p. 99.

6. *IBM Journal of Research and Development* 24.2 (1980).

7. Lewis M. Branscomb, "Electronics and Computers: An Overview," *Science* 215.4534 (Feb. 12, 1982): 759.

8. "King Research Publishes *Libraries, Publishers, and Photocopying,*" *Newsletter*, King Research, Inc. 3.2 (Oct. 1982):1. The character of electronic publishing is illustrated by the problem of citing the information in this paragraph, which came from these interest group exchanges themselves. Shall I cite the Arpanet list as from Zellich at Office-3?

9. 60 FCC 2d 261 (1976), Report and Order on "Resale and Shared Use of Common Carrier Services." In most countries this kind of brokerage or even shared use is not generally allowed. Most national telephone monopolies permit it only as an exception. In the United States it has come to be allowed as part of the general movement toward deregulation.

10. The data rate of a channel, which is expressed in "baud," can be described in terms of its bandwidth or conversely in the maximum number of bits per second that can be transmitted through it. Examples of familiar numbers are 300 bits per second for a slow-speed terminal, 48,000 bits per second for a voice telephone line, and 6 million bits per second for a television channel.

11. Compaine, *The Newspaper Industry*, p. 74.

12. Neustadt, Skall, and Hammer, "The Regulation of Electronic Publishing," pp. 375–378. In commenting on videotex, Richard E. Wiley, former chairman of the FCC, and Richard M. Neustadt remark: "This technology poses tricky policy questions. It combines publishing and broadcasting, and no one knows yet whether the FCC content regulations and other broadcasting rules will apply." Wiley and Neustadt, "U.S. Communications Policy in the New Decade," *Journal of Communication* 32.2 (1982): 30.

13. Jackson, Shooshan, and Wilson, *Newspapers and Videotex*, pp. 47–48.

14. Associated Press v. US, 326 US 1, 20, 89 L.ed. 2013, 65 S.Ct. 1416 (1945).

15. Lorain Journal Co. v. US, 342 US 143, 96 L.ed. 162, 72 S.Ct. 181 (1948).

16. Byars v. Bluff City News Co., Inc., 609 F 2d 843, 6th Cir. (1979), quoted in Neustadt, "The Regulation of Electronic Publishing," p. 379.

17. Joshua Lederberg, "Digital Communications and the Conduct of Science," *Proceedings of the IEEE* 66.11 (Nov. 1978): 1315–1317.

18. "Letters," *Science* 217 (Aug. 6, 1982): 217–218.

19. Encyclopedia Britannica Educational Corp. v. C. N. Crooks, 77–560, W.D. N.Y. (Feb. 27, 1978). See also Harriet L. Oler, "Copyright Law and the Fair Use of Visual Images," in John Shelton Lawrence and Bernard Timberg, *Fair Use and Free Inquiry*, (Norwood, N.J.: Ablex, 1980), pp. 268–286.

20. Some commentators have tried to limit First Amendment rights to matters of content and to distinguish these from property rights in plant. Nadel, "A Unified Theory;" Owen, "Economics and Freedom," p. 185. This is not in line with the historical tradition of the First Amendment. Records and files are protected under the Fourth Amendment, and printing plants are protected from licensing.

21. For FCC policy, see In the matter of Applications of Microwave Communications, Inc., 18 FCC 2d 953 (1969), 21 FCC 2d 19 (1970), 23 FCC 2d 202 (1970); MCI Telecommunications v. FCC, US, 41 RR 2d 191 D.C. Cir. (1971); Establishment of Policies and Procedures for Consideration of Applications to Provide Specialized Common Carrier Services, 24 FCC 2d 318 (1970), 29 FCC 2d 870 (1971), Docket 18920.

22. AT&T v. US, 572 F 2d 17, 2d Cir. (1978).

23. Docket 20828, Rules Adopted April 4, 1980.

24. Communications Satellite Act of 1962, USC Title 47, ch. 6.

25. USC Title 47, ch. 5, subch. 4, sec. 605.

26. 70 FCC 2d 1460, Docket 78–374, Oct. 18, 1979.

9. Policies for Freedom

1. For the changing technologies of communication, see Hiroshi Inose and John R. Pierce, *Information Technology and Civilization* (San Francisco: W.H. Freeman, forthcoming), chs. 1–2.

2. After the Italian Supreme Court ruled unconstitutional a monopoly in local radio and television broadcasting, hundreds of local stations were formed. The absence of a written constitution in Great Britain results in its approach being much like that of US Supreme Court Justice Frankfurter's, both viewing freedom as important, along with order, justice, and other values. But in Britain, without judicial review of constitutionality, it is left to the lawmakers to balance those values. Concern about the growing power of British trade unions to interfere with the content of newspapers and the mails, as well as about the effects of the Official Secrets Act, has generated a movement for the codification of rights. Unions in 1977 refused to deliver the mails of a struck plant and several times refused to print papers with ads or stories of which they disapproved. In Canada the presence of a controlling federal bill of rights was at the heart of the bitter resistance by Quebec to the Constitution instituted in 1982.

3. Similarly in the 1920s in Europe, labor governments for the first time took power and regulatory social legislation was adopted, but totalitarian movements arose to offer a populist counterattack to liberal reforms. Large

private enterprises came to dominate the press, and critics sought to control "irresponsible" media, particularly in the new technology of broadcasting.

4. The Marxist utopia is the freeing of all activity from resource constraints.

5. American newspapers are natural monopolies because they are so heavily advertiser supported; they would not be so if their readers paid most of the cost. The reason there are three broadcasting networks, not one, is the limit on the amount of advertising each broadcaster can carry. If the most successful network could add advertising minutes indefinitely, it would drive out the other networks. But as it is, there is too much advertising to fit on one or even two networks.

6. Miami Herald Publishing Co. v. Tornillo, 418 US 241, 41 L. ed. 2nd 730, 94 S.Ct. 2831. Lee Bollinger, "Freedom of the Press and Public Access: Toward a Theory of Partial Regulation of the Mass Media," *University of Michigan Law Review* 75 (1976): 1, argues that regulating some media and not others may be a good idea, even though the Supreme Court was wrong in basing broadcast regulation on a premise of special scarcity. This is a Frankfurtian brief for accepting the dangers of congressional experimentation.

7. In countries where cable television is operated by a government PTT rather than by private entrepreneurs, separation of carrier and content may make both economic and political sense.

8. Home Box Office, Inc., v. FCC, 567 F 2d 9, 185 US App. D.C. 142 (1977).

9. The Corporation for Public Broadcasting has recently been permitted a limited experiment running some ads. This barrier may be giving way.

10. Frost v. Railroad Commission of California, 271 US 583, 70 L.ed. 1101, 46 S.Ct. 605; Washington ex rel Stimson Lumber Co. v. Kuykendall, 275 US 207, 72 L.ed. 241, 48 S.Ct. 41. Cf. Stephenson v. Binford, 287 US 251, 77 L.ed. 288, 53 S.Ct. 181.

11. In Carterphone Dockets 16942 and 17073 of 1968, the FCC required phone companies to allow a radiophone service to be interconnected with them.

12. The FCC avoided confronting this issue by deferring action until the rates were about to be abandoned anyhow as the press moved over to the use of private lines.

13. Zechariah Chafee, *Free Speech in the United States* (Cambridge: Harvard University Press, 1941), p. 381.

Index